SP Pvt. Ltd.

Respectfully Dedicated to

DR. HARIDWAR SINGH

The Living Encyclopedia of High Energy Materials in India

Explosives Science

A Mathematical Treatise

Himanshu Shekhar

High Energy Materials Research Laboratory (HEMRL)
Defence Research and Development Organization (DRDO)
Ministry of Defence, Govt. of India
Sutarwadi Pune – 411 021

2013

Studium Press (India) Pvt. Ltd.

Explosives Science

A Mathematical Treatise

ISBN: 978-93-80012-43-8

Published by:

Studium Press (India) Pvt. Ltd.
4735/22, 2nd Floor, Prakash Deep Building
(Near Delhi Medical Association),
Ansari Road, Darya Ganj, New Delhi-110 002
Tel.: 91-11-43240200-15 (15 lines), Fax: 91-11-43240215;
E-mail: studiumpress@gmail.com

Printed at:
Akash Press, New Delhi 110 020

About the Author

Dr. Himanshu Shekhar is Joint Director at High Energy Materials Research Laboratory (HEMRL), Pune in India. He has Ph.D. in Mechanical Engineering. His GATE score was 99.57 and CPI of 10.00 out of 10.00, during M.Tech. from IIT, Kanpur. He joined Defence Research and Development Organization (DRDO) as Scientist 'B' and got outstanding promotions to rise to the level of Scientist 'F', now. He has contributed significantly in various projects of national importance and indigenous defence products developments.

He has authored 6 books on various aspects of high energy materials. His book on Rocketry in Hindi claimed Rajbhasa Pustak Puraskar – 2010 from DRDO. He has around 70 publications in peer reviewed journals and proceedings of international and national seminars and symposiums. He has delivered more than 25 invited talks and authored many popular science articles.

He is the recipient of "Agni Award for Excellence in Self – Reliance for the year 2001", "*Science Day Oration Award – 2003*", and "Young Scientist Award – 2004" from DRDO, India. He has been the first recipient of 'Mr. Engineer' Award in 2003 from the *Institution of Engineers (India),* Pune.

He has research interest in the area of high-energy materials including rocket/gun propellants and high explosives.

Preface

The Explosives are characterized by quick release of chemical energy, which has different mechanism of heat release than normal thermodynamic changes. The combination of chemical energy release and creation of physical shock wave makes action of explosives, a unique stream. The system design, initiation mechanisms, performance prediction, manufacturing and testing of designed systems leads to surprises on many occasions. Any new comer to the field of high explosive is generally detracted by the unpredictability in the behaviour of explosives.

In this context, this book compiles the basic calculation strategies for high explosives including manufacturing and prediction methodologies. The book also compiles various empirical formulations for the prediction of properties. For a new comer, this book gives the fundamental mathematical and theoretical knowledge to proceed further for research, development, production and use of high explosives. High explosive is a vast field and everything cannot be incorporated in a book to make it concise, apt, specific and adequate. Some of the exclusions for this book, which are elaborated exhaustively elsewhere in other books are – Synthesis of Explosive Molecules, Chemistry of Explosives, Selection criteria of Explosives, Characterization of Explosives, Ageing of Explosives, Detection of Explosives, Commercial Explosives, Primary Explosives, Nanomaterials in Explosive, Shaped Charges, Warheads, Safety etc.

Overall the book is designed in twelve chapters. First chapter is introduction of high explosives and method of identification and their normal usage and/or misuses are discussed, in brief. Second chapter discusses physics behind detonation including hydrodynamic aspects. Third chapter discusses salient properties of individual explosive molecules and their calculation from chemical formula, density and heat of formation. Chemical heat release, detonation temperature, oxygen balance, molecular weight etc are calculated for a set of explosives. Fourth chapter elaborates strategy for calculation of explosive properties for explosive formulations. Manufacturing aspects and initiation of explosives are covered in subsequent chapters. Determination of explosive properties by empirical methods is illustrated with examples in chapter seven. This chapter may be treated as the heart of the book. Eighth chapter has a brief about calculation of TNT equivalent of explosives, which is followed by measurement of explosive properties. Tenth chapter is concerned with calculations for different geometries of explosive charges. Eleventh chapter has discussion on anomalies in empirical and theoretical results for explosives. Twelfth chapter has future trends in calculation strategies for explosives.

Overall the book has following features:

- Journal references are placed at the bottom of the page for continuity in reading and not at the end of the chapter or book.
- Empirical and theoretical aspects related to the calculation of explosives are discussed.
- Full steps for calculations are depicted for their better understanding and wider use for other new/unknown molecules.
- Reference books for further reading are given at the end of the book.

Nobody can become master to design explosive devices, by reading a book on explosives but I hope that the book will serve researchers, paramilitary forces, police personnel, scientists, students, academic institutions and production agencies, associated with high explosives, in one form or the other.

Place: Pune, India **Dr. Himanshu Shekhar**
Date : 20.07.2012

List of Contents

1

Introduction to High Explosives

Explosion and Explosives

Compared to explosives, explosion is more general terms used frequently to denote any physical, chemical or nuclear blast. More than military forces, rising terrorists' incidences has exposed general public to the devastating effects of explosion. This exposure has raised many concerns and before understanding these events, a thorough treatment of explosion is mandatory. Explosion is a violent, sudden, noisy and terrifying release of energy. Although there are several beneficial uses of explosives, but it is destructive power of explosive, which is publicized the most.

Detonation is one of the unique phenomena, exhibited by explosives. Condensed explosives are general representative of explosives and they are invariably nitrated hydrocarbons. They exhibit detonation pressure of 300–400 kbar (30–40 GPa) and materials in the vicinity of detonation are subjected to high pressures for microseconds duration. Detonation shock moves with speed of the order of 3–10 kilometers per second. Overall the material goes to high compression states, which are not encountered in any other type of physical or thermo-mechanical systems. In fact, detonation can alter material surfaces and phase due to high pressure and small duration application in a unique way as cannot be obtained by other means.

Mostly, the explosives are used for military, mining and demolition applications, but it can also be used for material processing, precision cutting and pulsed power applications. As far as advantages of explosives are concerned, it is used for controlled blasting to remove hard rocks. Demolition of buildings, bridges and civil structures need explosives, customarily. The cleared space is utilized for making roads, dams, and buildings and in many cases, agricultural land is also cleared using explosives. In oil well exploration, the explosives are used for drilling and perforating hard rocks. The explosives are also used for control of avalanches and snow movement in high altitude rocky terrains. The civil explosive industry is flourishing due to these advantageous applications of explosives. The explosive forming, explosive cladding or explosive welding is, although not very popular as a commercial manufacturing venture, but due to quick and bulk material processing, potential utility exists in these areas. Detonation of explosive films are used in surface treatment and hardening of materials. Sintering, shock induced chemical synthesis, consolidation of powder are some other applications of explosives. For pulsed power applications, explosives are used in magnetic-flux compression, pulsed detonation engines, high intensity light pulse generator, etc. Explosives are also used for localized destruction of pathological tissues. It is also found that quickness of explosives makes them suitable for use in crashbags of cars.

Ammonium nitrate (AN), mainly used as soil fertilizer exhibits potential explosive properties. Despite being inferior to powerful military explosive, this is used as civil explosives along with mineral oil in the form of ANFO (Ammonium nitrate fuel oil). This is used in mining and has gained popularity due to excessive use by offenders, terrorists and separatists. Rather than using in pure form, fertilizers based on AN, is directly used to produce explosive effects. Low price, ease of access, high oxidizing capability, and compatibility with brisant explosives makes it ideal choice for exploitation in civil domain as potential explosive. Aluminium can be used as sensitizer of AN[1]. Ammonium Nitrate (AN) exhibits low heat of explosion (1592 kJ/kg against 4188 kJ/kg for an important conventional military explosive ingredient 2,4,6–trinitrotoluene

---

1 Bogdan Zygmunt, "Detonation Parameters of Mixtures Containing Ammonium Nitrate and Aluminium". *Central European Journal of Energetic Materials,* Vol. 6, No. 1, 2009, pp. 57–66.

or TNT). However, AN generates higher volume of gas (980 lit/kg against 740 lit/kg for TNT). Adding oil makes heat of explosion and volume of gas evolved for ANFO as 3890 kJ/kg and 975 lit/kg, respectively. Pure AN has density and detonation velocity of 0.909 g/cc and 3.560 km/s[2].

Despite many civil-applications, the criteria for explosive behaviour of a molecule have certain well-set rules. The first characteristic of an explosive is its sensitivity. The explosive molecules should be insensitive enough, not to get ignited during normal handling and should be sensitive enough to get ignited under suitable stimuli like heat, shock, impact, electricity, friction or other energy imparting mechanisms. The sensitivity and performance are two limiting factors, which govern behaviour and utility of the explosives. Considerable attention is given to introduction of trigger linkage in the explosive molecules and invariably presence of $C{-}NO_2$ or $N{-}NO_2$ makes molecules explosives. High-density is one of the prime requirements for explosives and same molecule, if used in propellants, exhibits high impulse. In fact, shockwave pressure, behind the detonation front is proportional to density square times specific impulse. This specific impulse depends on volume of gas produced and heat of combustion, both of which is a function of density again. So, overall dependence of detonation pressure on density is greater than quadratic. In addition to density, strain in the structure also imparts explosive behaviour in the molecule.

Another important aspect of an explosion is external environmental and operating condition. A substance may not explode at normal temperature and pressure but may show catastrophic behaviour at high temperature and pressure. Kinetic theory of gases justifies importance of temperature and promulgates essential condition for the occurrence of a chemical reaction as the number of collision of active species in favorable orientations with energy higher than activation energy. Increase in temperature of a chemical in gaseous form may cause increase

---

[2] Daniel Buczkowski, Bogdan Zygmunt, "Detonation Properties of Mixtures of Ammonium Nitrate based Fertilizers and Fuels". *Central European Journal of Energetic Materials,* Vol. 8, No. 2, 2011, pp. 99–106.

in number of atoms with energy higher than activation energy. This may increase reaction rate, which may lead towards explosion at elevated temperature readily. The pyrolysis, thermal dissociation, more formation of active species are also favorable conditions for high temperature explosions. However, pressure rise may also cause explosions in some of the substances, which are otherwise not explosives. Explanation on the basis of lowering of activation energy by application of high pressure cannot be very effective in explaining such behaviors. It was found that at a pressure of 3500 MPa, oxalic acid hydrate malonic, tartaric acid, citric acid, aspirin, sucrose, polystyrene, calcium chloride etc explodes on release of pressure. However such phenomenon is not observed with succinic acid, glutaric acid, adipic acid, maleic acid, fumaric acid and phthalic acid. It may be concluded that such materials needs even higher pressure to explode. This critical pressure above, which a substance explodes, may be dependent on coefficient of friction, plastic flow stress, and geometry of specimen. High pressure may cause internal heating and rise in temperature in some substances. However for explosive behavior such high pressure and temperature are not necessary and explosives should explode at very small energy input by suitable means.

Explosion is a process of rapid physical and chemical transformation of a system into mechanical work, accompanied with a change in its potential energy.

Classification of Explosives

The explosives are generally classified under many groups on the basis of chemical composition. First group is nitro aromatics and nitrobenzene as most obvious example. These molecules are characterized by polarizable π-electrons, resonance, inductive effects and conjugation. The more significant molcecules may have three or four nitro-groups. Sometimes substitutes are added to the benzene rings and it is observed that addition of $-OH$, $-NO_2$ group sensitizes the molecule, while $-NH_2$, $-CH_3$ etc have desensitizing effect. Nitramines are second class of explosives. Presence of lone pair of electrons in organic amines and aza-nitrogen weakens bonds. Examples are RDX, HMX, etc. The higher impact or shock sensitivity of nitramines as compared to nitro aromatics may be attributed to rotational energy of C–NO_2 bonds. This is between 0–5 kcal/mol for Nitro methane[3], nitro benzene etc but is higher than 9 kcal/mol for di-methyl nitramine. In general the rotational energy of X–NO_2 may lead to faster initiation of decomposition of molecules. The lower rotational energy of C–NO_2 in nitro aromtics than that in N–NO_2 of nitramines makes nitramines more sensitive. Higher N–NO_2 rotation energy impedes dissipation of thermal energy leading to higher sensitivity. Another type of explosives is characterized as strained molecules. Strain is present in a molecular entity and transition structures, if their energy increases due to unfavorable bond lengths, bond angles, dihedral angles relative to standards. Three, four or five member rings fused in a cage fashion are classified as strained cage-like molecules. Some salient examples are tetrahedrane, triprismane, cubane, homocubane, etc. Another class of high explosives, being developed falls in the category of high nitrogen content molecules. Nitrogen catenation is associated with instability and / or sensitivity. This is attributed to many factors like delocalization of charge, smaller N-N bond length, etc.

From the chemical point of view, way back in 1909, van't Hoff attempted to separate explosives from non-explosives on the basis of their chemical structure. As per the theories, chemicals containing following groups may be explosive molecules:

- Peroxides, ozone, ozonoids with O-O group
- Chlorates and perchlorates with O-Cl group
- Nitrogen chloride with N-Cl group
- Nitro-compounds, nitric acid esters and salts with N = O group
- Diazo-compounds, hydrazoic acid, its salts and esters with N = N group

---

3 Himanshu Shekhar, Explosive Characteristics and Shaped Charge Applications of Nitromethane (NM): A Review, Central European. *Journal of Energetic Materials*, Vol. 9, No. 1, 2012, pp. 87–97.

- Fulminates and cyanogens with N = C group
- Acetylene and acetylides with C ≡ C group

V. Plets, in 1953, modified the classifications of explosives on the basis of chemical structure. The explanation was in line with presence of certain specific groups in explosive molecules as suggested earlier by van't Hoff. However, Plets named such groups as *explosophores*. He proposed another set of groups called *auxoploses*, which are groups to protect, fortify or modify explosive properties in the molecules imparted by the explosophores. Plets' classification was based on the presence of explosophores in explosive molecules. The eight explosophores suggested by Plets are as follows

- $-NO_2$ and $-ONO_2$ in both inorganic and inorganic substances
- –N=N– and –N=N=N– in organic and inorganic azides
- $-NX_2$, where X can be halogen
- –N=C in fulminates
- $-OClO_2$ and $-OClO_3$ in inorganic and organic chlorates and perchlorates respectively.
- –O–O– and –O–O–O– in organic and inorganic peroxides and ozonides respectively
- –C≡C– in acetylene and metal acetylides
- M–C metal bonded with carbon in some organometallic compounds

The auxoploses do not make a chemical explosive, but may influence their explosive properties. Hydroxyl, carboxyl, chlorine, sulphur, ether, oxygen, amine etc are examples of such groups. In earlier days separating explosives from non-explosives are mainly based on these criteria, but later the role of oxygen balance and heat of explosion was also included for identification of explosive molecules. Researchers debated inclusion of these properties at length and it remained undecided, whether they were significant or not for identification of explosive molecules from their chemical structures. It was found that oxygen in carboxyl and hydroxyl may not contribute effectively in performance of a chemical as explosive because of their high heat of formation. Compared to them, the bonds with low heat of formation and low oxygen content (like N–O, C≡C) may affect explosive properties significantly.

The explosophores are referred as *plosophores* also and they are capable of forming explosive molecules, if introduced in hydrocarbons. The production of explosive power varies among these groups effectively and consistently. The classification as primary and secondary explosives is based on such power release mechanism. Primary explosives generally contain nitrate esters, aromatic and aliphatic nitro-groups and the nitramine groups. Secondary explosive has azo, azide, nitroso, peroxide, ozonoids, perchlorates etc. A set of hybrid explosives is also possible, if both the groups are present in a molecule.

Explosives are conventionally *classified as primary and secondary explosives.* Primary explosives are highly sensitive molecules, used in initiators. Mercury fulminate, lead azide, lead styphnate etc are common examples of such class of materials. The high sensitivity of these explosives may be attributed to their electronic conductivity. These are generally very good semiconductor materials, even at room temperatures. More than that, their photo-conducting behaviour sensitizes them by metallic impurities that are always present to promote their conductivity. They have myriads of metallic nuclei acting as active donor centers near the conduction band. Even a little decomposition quickly builds up the conductivity. At elevated temperature, release of electrons enhances conductivity further. The main mechanism of high sensitivity may be metallic speck and crystal imperfections. They have high densities and have detonation velocity in the range of 3 to 5 km/s. Another class of primary explosives may not be heavy metal salts like mannitol hexanitrate, tetrazine etc. These molecules have very high exothermicity due to presence of excitable electron donor centers. Presence of –N=N– group may promote release of electrons by photo or thermal means. These materials have very fast deflagration to detonation transition (DDT). Although primary explosives have high sensitivity, but now-a-days conformance to insensitive munition criteria demands use of less sensitive initiation compositions. More than use of various compounds, the initiation mechanism is made more safe by incorporating exploding bridge wire, exploding foil initiation, laser ignition, semiconductor bridge wire etc[4]. Primary explosives have low value of activation energy. This is representative of high reaction rate. When activation energy of a substance is zero, all molecules colliding during chemical reaction may react. As value of activation energy increases, this tendency to readily react, reduces. The pre-exponential factor for such explosives is also low. Overall this is responsible for requirement of less energy for initiation of primary explosives. This makes primary explosives more sensitive.

Mercury fulminate, lead azide, lead styphnate, lead mononitroresorcinate (LMNR), potassium dinitrobenzofurozan (KDNBF) etc are examples of primary explosives. Mercury fulminate [Hg $(C{\equiv}NO)_2$] is first reported by Wieland (1909) and Solonina (1910). It is sensitive to impact and friction and can be detonated by sparks and flames. It is very sensitive to sunlight and decomposes with evolution of gases. It has a crystal density of 4.42 g/cc and has molecular weight of 284.6. It decomposes at 90°C. Lead azide (PbN_6) was first prepared by Curtis in 1891. It is less sensitive to impact than mercury fulminate and more sensitive to friction. As it is less sensitive to impact, it is generally mixed with aluminium and lead styphnate to make ASA (Aluminium-Styphnate-Azide) mixtures for use in military detonators. It has molecular weight of 291.3 and decomposition temperature of 190°C. Lead styphnate [Pb $(HC_6{-}(NO_2)_3 - (OH)_2)$] is a weak primary explosive due to high percentage of metal (44.5%). It is an aromatic molecule with molecular weight of 468.3. It is not used alone in explosive detonators. It is sensitive to spark. Silver azide, Tetracene etc are also used as primary explosives.

Compared to primary explosives, secondary explosives are another class, which are relatively less sensitive, but they have high detonation velocities generally higher than 6 km/s. Common examples are TNT (2,4,6 trinitrotoluene), RDX

---

[4] Beat Vogelsanger, Beat Berger, "Energetic Materials – Performance and Safety", 36th International Annual Conference of ICT & 32nd International Pyrotechnics Seminar, Karlsruhe, Germany, June 29–July 1, 2005, Keynote Lecture, 0–1.

(Cyclotrimethylenetrinitramine), PETN (penta erythritol tetranitrate) etc. These materials exhibit low sensitivity because they are insulators with high electrical resistivity. In fact they are borderline primary explosives because DDT in this case is quite sluggish compared to primary explosives, but generates high temperatures to generate appreciable electron densities. These classess of explosive molecules are organic in nature with nitro groups (NO_2). Common nitro compounds have $C–NO_2$, while nitrate esters have $C–O–NO_2$ and nitramines have $C–N–NO_2$ groups. The secondary explosives have high value of activation energy and pre-exponential factor. This makes them more difficult to initiate than primary explosives and less sensitive to external stimuli.

2,4,6-Trinitrotoluene (TNT) $[H_2C_6–CH_3–(NO_2)_3]$ is most widely used military explosive. It has low melting point (80.65°C), low sensitivity to impact and friction. It is found in several isomeric forms, each having melting point higher than the 2,4,6-TNT. It is either colourless orthorhombic crystals or yellow monoclinic needles. It has specific gravity of 1.654. It is ignited at 300°C and heat of detonation is 1011 kcal/kg. 2,4,6-isomer is relatively insensitive to impact, friction, shock and electrostatic discharge. 2 kg drop hammer from 1.30 meter height can only ignite it. However presence of abrasive impurities and other isomers makes it slightly more sensitive to external stimuli.

1,3,5-trinitro-1,3,5-triazacyclohexane (RDX) $[H_6C_3–(NNO_2)_3]$ is also known as hexogen, cyclonite, cyclo-trimethylene trinitramine etc. it has high chemical stability and greater explosive power. It has melting point of 202°C and is available in white orthorhombic crystal forms. It gets ignited at 229°C and has heat of detonation of around 1324 kcal/kg. It is one of the most popular high explosive molecule and is used with TNT in many military explosives. It is also used as main filling as well as booster charges for initiation of main charges.

1,3,5,7-tetranitro-1,3,5,7-tetraazacyclooctane (HMX) $[H_8C_4–(N–NO_2)_4]$ has almost similar structure as RDX except for more number of carbon. It is also known as cyclotetarmethylene tetranitramine, octogen etc. It has more energy than RDX. It exists in many polymorphs – α (Orthorhombic), β (Monoclinic), γ (Monoclinic), δ (Hexagonal). The second form is least sensitive to impact and is most stable. Thus β-HMX is used mostly in military explosives. It has melting point in the range of 276–280°C and has heat of detonation of 1355 kcal/kg.

Pentaerythritol tetranitrate (PETN) $[C–(CH_2–ONO_2)_4]$ is a white crystalline (Tetragonal) explosive used in blasting caps, detonators, boosters[5], plastic explosives and other explosive formulations due to high detonation temperature and sensitivity. It is also known as Penta or Ten. It has melting point of 141.3°C. Due to symmetrical structure, it is chemically stable. It has heat of detonation of 1500 kcal/kg. Sensitivity of PETN is brought down by addition of wax as phlegmatizer. For a pressing density of 0.95 g/cc, the velocity of infinite diameter was calculated as 5.27 km/s and reaction zone length is found to be 0.54 mm[6].

Picric acid is a strong acid and is very toxic. Chemically, it is 2,4,6–trinitrophenol $[H_2C_6–(NO_2)_3–OH]$. It is yellow crystalline solid and has molecular weight of 229.1. It has density of 1.767 g/cc and has melting point of 122.5°C. It is superior to TNT in explosive properties. It is used in grenades and mine fillings. It is soluble in hot water, alcohol, ether, benzene and acetone. It attacks common metals and makes very explosive salts. However aluminium and tin are not attacked by this explosive.

Tetryl is also known as 2,4,6-trinitrophenylmethylnitramine $[H_2C_6–(NO_2)_3–N–NO_2–CH_3]$. It is frequently used in blasting caps. It is yellow in colour and has molecular weight of 287.1. It melts at 129.5°C and is moderately sensitive to initiation by friction and percussion. It is more sensitive than picric acid and TNT.

---

5 R.H. Dinegar, M.S. Millican, "The Initiation of Tetryl by Small Charges of PETN". *Propellants Explosives Pyrotechnics*, Vol. 1, No. 5/6, December 1976, pp. 111–113.

6 R.H. Dinegar, "Detonation Velocity of PETN in small Confined Cylindrical Charges". *Propellants Explosives Pyrotechnics*, Vol. 1, No. 5/6, December 1976, pp. 97–100.

Hexanitro stilbene (HNS) $[(CH)_2–(H_2C_6–(NO_2)_3)_2]$ is called heat resistant explosive and is less sensitive to initiation by impact, friction, spark or radiation. It is used as booster explosive. It is yellow crystalline solid with melting point of 318°C. It has molecular weight of 450.1 and has a crystal density of 1.74 g/cc.

TNT, RDX, HMX, Tetryl, PETN, HNS etc are main ingredients of explosive formulations. However attempts are on to replace them with suitable molecules. The main criteria for replacement are more power, less sensitivity, high density and high performance. 1,3,3–trinitroazetidine (TNAZ) is a possible replacement of TNT in melt-cast explosive. It is superior to TNT in performance and is more thermal stable and insensitive compared to RDX. Its decomposition in autocatalytic manner and presence of acid or base also catalyzes its decomposition[7]. It $[NO_2–N–(CH_2)_2–C–(NO_2)_2]$ has molecular weight of 192.08 and has crystal density of 1.84 g/cc. It has melting point of 101°C and has a boiling point of 252°C.

Advanced Explosives

CL-20 or hexanitrohexaaza isowurtizitane (HNIW) is considered one of the most powerful members of high energy density materials (HEDM) class of explosives. It has high density (> 2g/cc) and positive heat of formation (~ 419 kJ/mol), due to cage structure, presence of six nitro groups and superior oxygen balance (–11%). Compared to this HMX has a density of 1.91 g/cc, heat of formation of 71 kJ/mol, only 4 nitro groups and oxygen balance of –22%. This is reflected in explosive performance of the molecules also. It has a velocity of detonation of 9429 m/s (HMX has 9162 m/s) and detonation pressure of 43.55 GPa (HMX has 39.55 GPa). In plastic bonded explosives (PBX) based on Polyurethane (PU) binder, practically CL–20 based composition exhibits higher density and velocity of detonation. 95/5 composition of explosive/PU gives density of 1.75 g/cc for HMX composition while for CL–20 based formulations, it is upto 1.83 g/cc. At these densities, velocity of

---

[7] V.V. Nedelko, B.L. Korosounskii, N.V. Chukanow, T.S. Larikova, N.N. Makhava, I.V. Ovchinnikov, "Thermal Decomposition of 1,3,3-Trinitroazetidine in Gas Solution and Melt", 37th International Conference of ICT, Karlsruhe, Germany, June 27–30, 2006, pp. 154–1.

detonation of HMX and CL–20 based formulations are practically determined as 8533 m/s and 8971 m/s respectively[8]. CL–20 is considered powerful and sensitive explosive and its formulations also depict similar behaviour. Run to detonation distances are very small for CL–20 based formulations and this does not change significantly with pressure. Compositions based on RDX, HMX show pressure dependent run to detonation distances[9]. Acceleration ability of CL–20 is experimentally found to be 7% higher than that of HMX. Addition of aluminium powder increases heat of detonation and acceleration ability of CL–20[10].

Highly nitrated cubanes are explosives of future. They are predicted to have very high density and very powerful explosive properties. High nitrogen content and highly strained cubic cage structure increases heat of formation with addition of nitro groups. The normal cubane has experimental heat of formation in gas phase as 621.6 kJ/mol, where as octanitrocubane (ONC) can have calculated heat of formation in the range of 730 kJ/mol to 950 kJ/mol (by different approaches). The velocity of detonation and detonation pressure of octanitrocubane is reported to be 10.05 km/s and 48.47 GPa respectively[11]. The general formula for polynitrocubane is $C_8H_{8-m}(NO_2)_m$ and as per isodesmic reaction with molecules of methane (mCH_4), it gives cubane (C_8H_8) and methyl nitrate (mCH_3NO_2). Heats of formation of methane, cubane and methyl nitrate are –74.6 kJ/mol, 621.6 kJ/mol and –74.3 kJ/mol respectively. More than five substitutions of hydrogen by nitro-groups result in improvement in explosive properties due to raised

---

8 Samson S. Samudre, U.Shadevi R. Nair, Girish M. Gore, Rabindra Kumar Sinha, Arun Kanti Sikder, Shri Nandan Asthana, "Studies on an Improved Plastic Bonded explosive (PBX) for Shaped Charges". *Propellants Explosives Pyrotechnics*, Vol. 34, No. 2, 2009, pp. 145–150.

9 Thomas Eich, Roland Wild, "Run to Detonation Distance of Phlegmatized CL–20 Compared to Other High Explosives", 37th International Annual Conference of ICT, Karlsruhe, Germany, 27–30 June 2006, pp. 52–1.

10 Michael N. Makhov, "Evaluation of the Epxlosive Performance of Cl–20", 37th International Conference of ICT, Karlsruhe, Germany, June 27–30, 2006, pp. 100–1.

11 Xue-Hai Ju, Zun-Yao Wang, "Theoretical Study on Thermodynamic and Detonation Properties of Polynitrocubanes". *Propellants Epxlosives Pyrotechnics*, Vol. 34, No. 2, 2009, pp. 106–109.

density and heat of formation. Addition of nitro group gets jammed into cubic skeleton and improves detonation velocity and pressure both. Another compound reported to have very good explosive properties is 1,3,5,7-Tetranitro-2,4,6,8-Tetraazacubane[12]. It has density of 1.94 g/cc and heat of formation of 218 kcal/mol. The detonation velocity detonation pressure is reported to be 9.80 km/s and 44.4 GPa respectively.

Several high-nitrogen high-oxygen explosives are also reported as a possible replacement of RDX with matching of superior density and velocity of detonation. In this category of compounds, 5-nitriminotetrazole (H_2AtNO_2), 5-aminotetrazolium dinitramide (HAtDN), N^3N^6–bis (2,2,2–trinitroethyl)–1,2,4,5–tetrazine–3,6–diamine (BiTNEAT), anhydrous bistetrazolylamine (H2BTA), triaminoguanidinium dinitramide (TAG-DN) and triamino-guanidinium 1-methyl-5-nitrominotetrazolate (TAG–1–$MeAtNO_2$) are reported[13], which showed theoretical velocity of detonation of 9450 m/s (1.87 g/cm^3), 9429 m/s (1.86 g/cm^3), 9261 m/s (1.89 g/cm^3), 9120 m/s (1.86 g/cm^3), 8796 m/s (1.628 g/cm^3) and 8765 m/s (1.57 g/cm^3) respectively as against 8936 m/s (1.82 g/cm^3) for RDX.

It is not high energy or power of explosives, which are driving force for synthesis or exploration of new molecules, but sometimes sensitivity is also a governing factor. Various methods are adopted for reduction in sensitivity of secondary explosives.

- Change crystal properties of explosive molecules (I-RDX, RS-RDX).
- Use of insensitive high explosive (IHE) molecules in melt-cast formulations.

---

[12] Peter Politzer, Pat Lane, Jane S. Murray, "Computational Characterization of a Potential Energetic Compound: 1,3,5,7-Tetranitro- 2,4,6,8,-Tetraazacubane", *Central European Journal of Energetic Materials*, Vol. 8, No. 1, 2011, pp. 39–42.

[13] Thomas M. Klapotke, Michael Gobel, Jorg Stierstorfer, "High-Nitrogen and High-Oxygen Epxlosives as Possible Replacement for RDX", *International Journal of Energetic Materials and Chemical Propulsion*, Vol. 10, No. 1, 2011, pp. 45–54.

- Embedding explosives in elastic binders matrix (PBX).
- Coating of Insensitive explosives on conventional explosives.

FOX-7 or DANDE or 1,1-Diamino-2,2-dinitroethene is one such explosive with low sensitivity and high performance. The unique characteristic is imparted by layered structure with strong intermolecular hydrogen bonds within layers. This was synthesized in 1998. For impact sensitivity test with 5 kg hammer, impact height is found 23 cm against 6 cm for RDX. This corresponds to impact energy of 11.3 J against 2.9 J for RDX. Height for 50% explosion is found as 35.2 cm which corresponds to impact energy of 17.3 J, while for HMX, this energy was 9.6 J. Friction sensitivity of FOX-7 was measured as 353 N by Julius-Peters Machine. The measured detonation velocity is 8325 ± 80 m/s (Theoretical 8453 m/s by BKWC). The average heat of detonation is 4860 ± 60 J/g (Theoretical by BKWC 4774 J/g). Cylinder test results shows equivalence of FOX-7 with phlegmatized (95%) RDX[14].

On the low sensitivity front, 1,3,5-Triamino-2,4,6-Trinitrobenzene (TATB) $[C_6–(NO_2)_3–(NH_2)_3]$ is one of the promising explosive molecule with density of 1.94 g/cc and velocity of detonation of around 8000 m/s. Methods of its synthesis by various methods is discussed in a paper, where Benziger synthesis, Atkin synthesis, Synthesis from Phloroglucinal, VNS chemistry, synthesis from Pycramide etc are described and compared[15]. Its ignition temperature on progressive heating is above 300°C. One of the distinguishing features of TATB (resembling that for FOX-7) is observed in differential thermal analysis (DTA). When heated at heating rate higher than 5°C/min, single exothermic peak is observed, while at slow rate of heating two exothermic peaks are

---

[14] Waldemar A. Trzcinski, Stanislaw Cudzilo, Zbigniew Chylek, Leszek Szymanczyk, "Investigation of Sensitivity and Detonation Properties of FOX-7, 37th International Conference of ICT, Karlsruhe, Germany, June 27–30, 2006, pp. 57–1.

[15] Alexander R. Mitchill, Philip F. Pagoria, Robert D. Schmidt, Michael D. Coburn, Gregory S. Lee, Peter C. Hsu, "A Versatile Syhthesis pf 1,3,5-Triamino-2,4,6-Trinitrobenzene (TATB)", 37th International Conference of ICT, Karlsruhe, Germany, June 27–30, 2006, pp. 137–1.

observed for TATB, indicating a 2–stage decomposition mechanism. The presence of two peaks is seen in recrystalized TATB also[16].

NTO (3-nitro-1,2,4-triazole-5-one) [O=C–$(NH)_2$–(C–NO_2)–N] is another insensitive cxplosive molecules. It has high heat of reaction and is less sensitive and more stable than RDX. It is white in colour and has a molecular weigh of 130.1. It has density of 1.93 g/cc and has decomposition temperature of 273°C. NTO/TNT based formulations are being investigated for its properties and safety parameters. Attempts also made to coat HMX with NTO in a solvent solution by their subsequent evaporation. Growth rate of coating and surface morphology is altered suitably to achieve uniformity and adequacy of coating[17].

Another molecules 5–Oxo–3–Nitro–1,2,4–triazole (ONTA) is also reported to be a molecule of insensitive characteristics with density of 1.91 and velocity of detonation of 8590 m/s. Pressed PBX compositions are formulated using HMX/TATB and HMX/ONTA from moulding powder using Kel-F plastic binder (7%) and graphite (0.5%) as lubricating agent. Compared to pure HMX composition, these ingredients imparted high insensitivity to friction (from 210 N to above 300 N) and impact (from 3.6 J to round 10 J)[18].

The military explosives are never used as single ingredient or molecule. Invariably properties are tailored by combination of materials. Sometimes manufacturing ease also plays a vital role in preparation of explosive formulations. Sensitivity, energy content, velocity of detonation, detonation pressure, etc are major driving force for development of different classes of compositions. AMATOL is a mixture of TNT and AN. If Picrite replaces AN, the

---

16 V.P. Filin, N.V. Garmasheva, I.V. Chemagina, N.P. Taybinov, V.T. Timofeev, "Features of Thermal Stability of TATB Monocrystals", 37th International Conference of ICT, Karlsruhe, Germany, June 27–30, 2006, pp. 97–1.

17 Kwang-Joo Kim, Hyoun-Soo Kim, "Coating of NTO on HMX using Surface Crystallization", 36th International Annual Conference of ICT & 32nd International Pyrotechnics Seminar, Karlsruhe, Germany, June 29–July 1, 2005, pp. 187–1.

18 Alain Becuwe, Alain Delclos, "Oxynitrotriazole and its use as Insensitive Explosive", Proceeding of the International Symposium on Pyrotechnics and Explosives" Beijing, China, Oct 12–15, 1987, pp. 255–261.

composition is called PICRATOL. AMMONAL (MINOL) is a mixture of AMATOL and aluminium.

CYCLOTOL is mixture of RDX and TNT. If RDX of CYCLOTOL is replaced with HMX, the composition is called OCTOL. If RDX is replaced with PETN, the composition is called PENTOLITE. If RDX of CYCLOTOL is replaced with tetryl, the composition is called TETRYTOL. Aluminized CYCLOTOL is called TORPEX. Sometimes Aluminium is loaded in TNT to give composition called Tritonal.

Another class of explosive formulations is named as Composition followed by English alphabets. Composition–A–3 is mixture of RDX and wax. Composition-B is combination of RDX, TNT and wax. Composition-C is RDX with plasticizer and/or TNT/ DNT/MNT etc. A brief about various types of explosive ingredients presently in use and their future for military applications were enumerated in a paper[19].

Mathematics with Explosive Behaviour

Explosives when suitably initiated can show its devastations in the form of thermal, blast, fragmentation, shattering, compression, projection and penetration effects. The explosive molecules are generally characterized by density and heat of formation along with their chemical structure.

Explosives are different from other ordinary fuels (higher electro-negativity) because they carry their own oxygen. Explosives do not possess higher energy content, but rate of reaction is important and explosives have very high rate of chemical reaction.

> *Explosives are thermodynamically unstable systems, which under the influence of external action (stimuli like spark, impact, friction, shock, heat etc) are capable of extremely rapid exothermic transformations, accompanied by the formation of very heated (exothermic chemical reaction or production of heat) gases and vapours.*

[19] Jörg Mathieu and Hans Stucki, Military High Explosives, CHIMIA, Vol. 58, No. 6, 2004, pp. 383–389.

There are *three essential requirements* of detonation through chemical explosion –

- chemical reaction should propagate faster than bulk speed of sound,
- gases should be produced after chemical reaction and
- heat should be produced as a result of chemical reaction.

Chemical reaction augments shock wave in detonation front and maintains a constant velocity of detonation. The rate of reaction is very high. As far as order of reaction is concerned, there is no unique observation in this respect. Exact mechanism for decomposition of explosives is not known, but most of the explosives may be exhibiting zero order reaction. So rate of reaction is independent of composition or presence of reaction products or active species. It may have some dependence on temperature in the form of Arrhenius equation. However, the temperatures, which are encountered in chemical explosion, are so high that zero order reaction kinetics are not valid. Type of explosives and extent of decomposition affects order of reaction and impurities are also found to catalyze the chemical reaction. AN, TNT, RDX, Tetryl and PETN are reported to decompose according to first order decomposition law. It is also mentioned that reaction in isothermal geometrical mode may have an order of 2/3. Overall detonation reaction is pseudo-2/3 order but intrinsically first order.

Volume of gas produced per unit mass of explosive is generally of the order of 500 to 1000 cc per gram of explosive. However in the thermite reaction (iron oxide Fe_3O_4 and aluminium), no gaseous products are formed but it may lead to explosion due to tremendous amount of heat liberated. The temperature of the order of 2500°C may be attained during this process. It may be noted that detonation products are all solid during thermite reaction. Generally explosive should produce higher quantity of gaseous molecules. In other words, in explosion reaction number of moles of products should be higher than number of moles of reactants. However formation of water from hydrogen and oxygen ($2H_2 + O_2 \rightarrow 2H_2O$) is an explosive reaction where reduction in number of moles occurs on chemical reaction. The explosive nature is imparted by tremendous amount of heat released during the chemical reaction. The reduction in volume is compensated by high exothermicity and high rate of process, which generates very high pressure after this reaction.

The explosion or detonation is an exothermic reaction and heat produced is in the range of 2500 to 7000 kJ/kg (600 to 1800 cal/g). However there are explosives with endothermic bonds, like polyacetylenes. Although their bonds break in endothermic manner but even then they are explosives due to high volume of gases produced and high rate of chemical reaction. The initiation of explosive should be spontaneous (chemical reaction should move towards formation of gaseous product-molecules) and self propagating (no external supply of energy needed after initiation). Energy produced per unit weight of explosive is almost same or even less than that for a normal fuel like petrol, coal, etc. However, explosive possesses high energy density, which is responsible for the greater power of explosives and their destruction ability.

The physics of detonation is one of the important aspects during explosion. Hydro-dynamic theories are developed and normal principles of detonics are discussed without any discussion on chemical reaction. The discussion on conservation equation of mass, momentum and energy along with equation of state for gaseous products of explosion are responsible for development of physical nature of detonation waves in a non-reacting solid. However the poor augmentation of such propagating shock wave through non-reacting solid (condensed phase) has decaying velocity, but real situation is entirely different, where these detonation waves are augmented by heat released in the chemical reaction. When shock wave is passed through any solid explosive, the explosive is assumed to get converted into reaction product in same volume instantaneously. Additionally shock wave leads to compression of explosives also. The detonation front moves in the explosives at high velocity and chemical reaction zone follows immediately behind it. The detonation front is observed as a discontinuity, which is often called von-Neumann peak. The reaction zone is followed by rarefaction zone and the demarcation of chemical reaction zone and rarefaction zone is called Chapman-Joguet point. The physics of detonation discusses these aspects of the solid explosives.

The essential properties of explosive molecules are derived from their chemical structure, density and heat of formation. Positive heat of formation, higher density and positive oxygen balance are main requirements for a substance to qualify for explosive. From these basic input parameters, volume of gas evolved, molecular weight, oxygen balance, detonation temperature from single explosive molecules can be calculated. The properties of single explosives discuss these aspects in general. Since end effects desire several properties, which cannot be fulfilled by single explosive molecules and sometimes combination of explosive molecules are used. The use of AMMOTOL, BARATOL, TORPEX, H-6 COMPOSITION-A3, COMPOSITION-B, COMPOSTION-C etc are examples of some of the explosive compositions, which are developed for filling in special bombs, grenades, warheads or mines. The properties of such explosive compositions demand addition mathematical calculations and invariably such combinations are converted into equivalent explosive molecules with properties dependent on the mixed explosive molecules. The mathematical treatment of such explosive combination is interesting and different from tackling single explosives.

Explosives for different uses are generally manufactured by pressing, casting, rolling, extrusion etc. The percentage ingredients of different ingredients of explosives in the final composition desire many calculations. The pressing demand calculations related to mold design, length of mold and plunger etc. The casting desire not only selection of explosive ingredients but process parameters are also important. The casting is generally executed as melt-cast or chemical cross-linking. The rolling also depend on chemical cross-linking. During manufacturing of explosives, a major challenge is to keep watch on sensitivity of explosives and invariable non-availability of explosive molecules in pure form desire many calculations. The pure explosives are generally available in water or with other desensitizer or phlegmatizers. Making correct concentration in final composition needs mathematical tools for compensation of such extra ingredients.

The initiation of explosives is another area, which needs sensitivity assessment as well as ease of initiation of explosives.

The hot spot theories, initiation mechanisms, thermal runway conditions, effects of other stimuli on condensed phase etc needs mathematical theories. The initiation of explosives is followed by performance of explosives in terms of detonation parameters like detonation velocity, detonation pressure, ratio of explosive heats etc. The calculation of these parameters through many softwares like RUBY, TIGER, CHEETAH, BKW etc are well reported and role of researchers using these codes becomes restricted to mere feeding data and getting end results. The calculation of detonation parameters by empirical relations is well researched and many methods for hand-calculations of detonation parameters of explosives are discussed. Obviously they demand mathematical formulations and statistical tools. Using these empirical tools, calculation of detonation parameters become very simple using a hand-held calculator.

Comparison of various explosive formulations and molecules on the basis of their TNT equivalent is also a proven method, where sand-crush test, air-blast test, ballistic mortar test etc are used. In all the methods some parameters of new explosive is measured along with performance of TNT and TNT equivalent is determined as amount of TNT needed to produce same effect as produced by unit mass of the tested explosive. Different method lead to different TNT equivalent and collection of all the parameters for calculation of TNT equivalent and resolving various discrepancies in the value of TNT equivalent for given explosives is a tedious and vague at present. The measurement of explosive parameters is a cumbersome task due to rapidity of action and also due to extreme values realized of different parameters. The explosive phenomena have very large time scales and in general, it becomes a transient process. Measurement methods for different properties also desire calculations of one form or the other.

Although mathematics is important for all aspects of explosives including properties, manufacturing, performance, comparison and end-effects, but mathematics is a tool only. Several effects are observed in real life which sometimes does need separate explanations. Such anomalies need discussion on case-to-case basis. The geometry of final shapes of explosive devices can vary and determination of explosive quantity invariably needs calculation of volume. The future directions in all the areas of explosive development do need simultaneous development of related mathematical theories. With time understanding explosive behaviour through mathematics is becoming more and more complex and an attempt to discuss mathematical aspects with explosion and explosive is worth investigation.

2

Physics Behind Detonation

Origin of Detonation

The detonation associated with chemical explosion of materials is three-dimensional in nature and has a number of associated parameters. It is the most intense and the fastest of all the chemical reactions. The total time associated with detonation may be of the order of microseconds, depending on physical and chemical nature of the explosives. Many high explosives have it in this range while ANFO (ammonium nitrate and fuel oil) and other civil explosives have total reaction time from 50 to 100 microseconds. Despite quickness, which can deter an amateur explosive researcher, these detonation events follow thermodynamic and hydrodynamic principles. One of the most remarkable observation about the detonation from high explosives is that detonation velocity is constant in all uniform explosives after propagating for a sufficiently long distance (called run-up distance) from the initiator. The sufficient length may be 3-4 times diameter of propagation.

The origin of shock wave may be traced to stress-strain behaviour of the materials. Under uniaxial compressive testing, stress is proportional to strain for smaller strains. During this period material shows elastic properties also and any reversal of stress causes disappearance of generated strain or deformation. During this period, sound velocity in the material is constant (c is proportional to change in Pressure/density or P/ρ). If above relation is relied upon, pressure and density are linearly related in elastic region for all materials. Any change in pressure brings proportionate change in the density of the material. Contrary to this, if materials are loaded at much higher stresses, rise in pressure surpasses density variations and sonic velocity rises. In this region material behaves as plastic material. This occurs at a much high value of stresses. There exists a region between (end of) elastic limit and (start of) plastic limit, where the material shows combined plastic and elastic behaviors. In study or understanding of shock the plastic limits are more important. When wave propagation occurs through a material in highly plastic stage, many velocity terms evolve. First is particle velocity (u), which is velocity at which the material is locally accelerated. When a pressure wave traverse at high pressure through a material, low pressure region lies in the front and high-pressure regions are generally trailing. However, at high pressure, particle velocity is high and so is sonic velocity. This indirectly indicates that pressure wave of high-pressure region moves at a faster pace than same of low-pressure regions. When pressure wave of high-pressure region catches pressure wave of low-pressure region, the pressure wave front becomes almost vertical, indicating a discontinuity called shock wave. The material can be treated as undergone from a state of unshocked to shocked state through this jump called shock wave.

One of the basic postulates called Chapman-Jouget (C-J) postulate is that detonation velocity (D) is sum of particle velocity (for gases cloud velocity) (u) and sonic velocity (c) in the following medium or $D = u + c$. The role of molecular structure of high explosive molecules is represented by equation of state (EOS), where co-volume plays a vital role due to high temperature and pressure conditions prevailing in the explosion clouds. Co-volume (η) is generally a function of pressure and volume both and equation of state may be written as $pv = nRT + \eta(T,v) \times p$. however, co-volume for the explosives is found to be independent of temperature and dependence on volume is only mathematically represented by many researchers. One of the expressions is $\eta = e^{-0.4\times\rho}$. Pressure versus density curves for most of the explosive follow an exponentially rising trend. Pressure is defined as energy density and in the context of explosives it is kinetic energy density. If all the chemical energy per unit volume of a given explosive is converted in detonation to energy density, then maximum possible pressure is realized. During detonation, chances of creation of plasma also exist.

Shock Wave Theory

Essential conditions for the generation of shock wave are stressing material beyond plastic limit in compression mode. A pressure wave travels through this compressed material. This results in increase of pressure wave velocity, finally resulting in a state of shock. Rarefaction waves move behind the shock front at a faster rate and shock is attenuated or neutralized by this. It is a chemical reaction driven shock wave in molecularly premixed material called explosive. The energy is released in the reaction zone behind the shock. This release of energy is converted into kinetic energy However, for ease of understanding; simple one dimensional steady state understanding behind physics of detonation can give a better insight in the phenomena and can also provide substantial input to the three-dimensional treatment of detonation waves. Detonation is generally characterized by shock waves, which propagates through gaseous medium at a very high velocity. Any continuous adiabatic flow through a pipe may result in acceleration of flow and ultimately giving a discontinuity. The entire phenomena of steady shock transmission through a medium may be shown in Fig. 1.

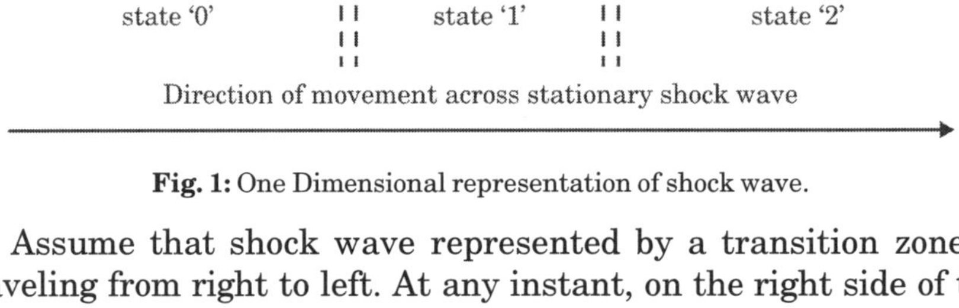

Fig. 1: One Dimensional representation of shock wave.

Assume that shock wave represented by a transition zone is traveling from right to left. At any instant, on the right side of the transition zone, there exists a zone of final state of the medium, which is attained due to effect of shock wave on the medium. Thermodynamic parameters will change due to movement of shock wave through the medium. Contrary to this medium to the left of the transition zone will have materials in their initial undisturbed state. The same situation can be understood by assuming steady state condition of shock wave and assuming that medium is moving across the stationary transition zone of the shock wave from left to right. Both the situations are depicted in Figure 1 and later method of assessment is better numerically. In fact, in true sense, the situation is not depicted, as a closed thermodynamic system and application of conservation equation demand certain assumptions. For the open system, balance of flux across each boundary may be conserved under assumption of control volume. The undisturbed state of the medium (left of transition zone) is depicted by subscript '0', while final state (right of transition zone) is shown by subscript '2'. The transition zone is depicted by subscript '1'. Mass, momentum and energy conservation equations are applied for the situation in hand.

Mass balance equation is also called continuity equation and states that mass flux density given by product of density and velocity is same everywhere in the flow. Neither spatial nor temporal variation and/or fluctuation is observed in the flow. The flow is steady and there is no accumulation or depletion of mass during the flow. The momentum balance equation is basically derived from equation of motion. It uses two parameters – one is momentum flux given by product of density and square of the velocity and other is pressure. Sum of pressure and momentum flux is constant and any reduction in pressure will always be resulting in a rise in momentum flux. Viscous dissipation terms also play a role in momentum equation, but it is generally neglected for ideal flow through shock waves. The energy conservation equation is valid for fluid flow. It states that change in total energy (kinetic energy + internal energy) is equal to sum of the work done by the forces acting on the fluid element and energy flux passing boundaries of the volume.

Although detonation wave traverses through solids, but its effect is predominantly seen in fluid state, mainly in gaseous form.

Gaseous state of matter under thermodynamic equilibrium is defined completely by two independent variables. This is true for any homogenous matter in thermodynamic equilibrium and since state is gaseous, principles of gas-dynamics describe the detonation events completely. Although properties of matter are classified as intensive (independent of mass) and extensive (dependent on mass), there is operational classification of thermodynamic properties of systems. One characteristic quantity is called mechanical quantity, which includes pressure, density, enthalpy, sonic velocity, internal energy, etc. Another class of quantities is called non-mechanical or thermo-caloric quantity.

Although a mixture of ingredients executes the detonation, but a certain gross composition can be ascertained, whose components are in constant chemical-thermodynamic contact. In fact explosives and its reaction product represent the same matter in two different states. Thermodynamic behaviour of matter is expressed in terms of equation of state, which is a relation between three state variables. The equation of state may be complete or incomplete depending on definition of thermodynamic state of matter described by it. Incomplete equation of state is described in the form $f(p,v,T) = 0$. This is called thermal equation of state. Most peculiar thermal equation of state is valid for thermally perfect gas and is written as $pv = RT$. For thermally perfect gas, specific internal energy and specific enthalpy is a function of temperature alone. $E = \int C_v(T)\,dT$ and $H = \int C_p(T)\,dT$. If, in addition to being thermally perfect gas, specific heat at constant pressure for the gas is constant (C_p = constant), the gas is called calorically perfect gas. Another way to define calorically perfect gas is by imposing ratio of specific heat as constant. $E = C_vT$ and $H = C_pT$ or $E = pv/(\gamma - 1)$ and $H = \gamma pv/(\gamma - 1)$. Since pressure volume is involved, last two set of equations may also be called mechanical equation of state for a calorically perfect gas. This equation is sufficient to determine isothermal curves on a p–v plane, but is insufficient to determine state variables like internal energy, specific enthalpy or specific heat at constant pressure. Supplementary equation of state called caloric equation of state of the form $C_p(p, T)$ or $C_p(v, T)$ is needed to get such parameters. Similarly another incomplete equation of state in gas dynamics is $E = E(p, v)$. This is called mechanical equation of state.

Complete equation of state is written for a gaseous state of matter in terms of internal energy (E), enthalpy (H), Gibb's free energy (G) and Helmholtz energy (F) and has four equations all together.

$$E = E(S, v); H = H(S, p); F = F(T, v); G = G(T, p)$$

The above-mentioned equations are called characteristic functions of state and each term is defined in terms of state variables.

$$H = E - pv; F = E - TS; G = H - TS$$

The above relations are valid for non-reacting mixtures and in this chapter, chemical reaction is completely dispensed with. However, chemical reaction may occur in adiabatic or nonadiabatic manner depending on whether heat generated during reaction (generally combustion or detonation) is contained within the system or dissipated outside. If entropy change is observed the reaction becomes irreversible. Sound wave and sock waves, both exhibit adiabatic process. Sound wave produces isentropic change of state, but flow through a shock wave cause entropy change.

Rayleigh Line

For explanation of shock propagation through a medium, it is assumed that medium is moving at certain velocity (u) with respect to the shock, which is also moving with certain high velocity (D). Density is represented by ρ, V represents volume and pressure is represented by p. The properties of free medium is depicted by subscript '0', while that for medium after passing through shockwave is represented by subscript '2'. This is in accordance with situation depicted in Fig. 1.

Conservation of mass:

$$\rho_0 D = \rho_2(D - u) \text{ or, } u = (1 - \rho_0/\rho_2)D$$

Conservation of momentum:

$$p_2 - p_0 = \rho_0 Du = \rho_0(1 - \rho_0/\rho_2)D^2$$ {from mass conservation equation}

or, $$p_2 - p_0 = \rho_0{}^2 D_2(1/\rho_0 - 1/\rho_2) = \rho_0{}^2 D_2(V_0 - V_2)$$

or, $$R = \rho_0{}^2 D_2 - (p_2 - p_0)/(V_0 - V_2) = 0$$

The last expression is called equation of Rayleigh line. It is derived from conservation of mass and momentum. Energy need not be conserved along Rayleign line. Rayleigh lines can be plotted on p–V plane for constant shock velocity (D). The initial state can be depicted by state p_0 and V_0. Slope of the Rayleigh line at constant shock velocity can be given by $(p_2 - p_0)/(V_2 - V_0)$, which is equal to $-\rho_0{}^2D_2$. When shock velocity is zero, ie there is no shock wave propagation through the medium; the initial pressure of medium is maintained till end. Slope of the curve is zero. Due to movement of transient domain, volume may change. When shock velocity is infinity ($D \rightarrow \propto$), it is called constant volume detonation. There is no change in volume and pressure rises tremendously and instantaneously. This has infinite slope. For any intermediate value of shock wave velocity, the curve is a straight line with negative slope emanating from the initial state. The pictorial representation of Rayleigh line is shown in Fig. 2. High value of shock velocity represents higher slope or steep variation and for same volume change, larger change in pressure is encountered.

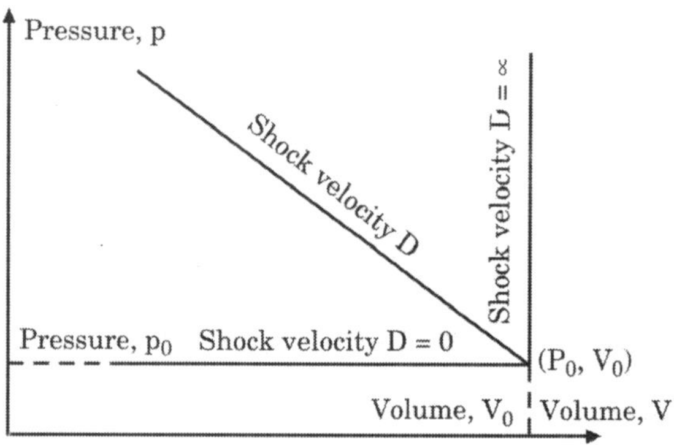

Fig. 2: Rayleigh line for constant shock velocity.

The combination of mass and momentum conservation equation can be rearranged by eliminating shock velocity (D) from both the expressions.

$$p_2 - p_0 = \rho_0 Du = u^2/(1/\rho_0 - 1/\rho_2) = u^2/(V_0 - V_2)$$

or

$$u^2 = (p_2 - p_0) \times (V_0 - V_2)$$

So, curves of constant medium velocity (u) are represented by rectangular hyperbola. Higher value of velocity of the medium makes curve move away from the starting point. Velocity of medium is also called particle velocity and the same is depicted in Fig. 3. For a given shock velocity, particle velocity and initial state, final state is given by intersection of Rayleigh line and rectangular hyperbola.

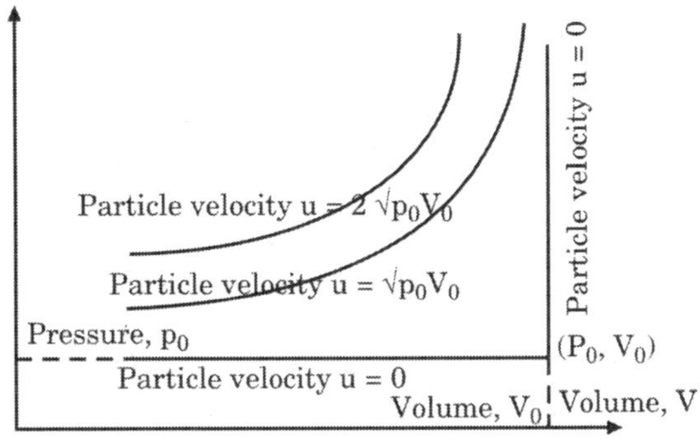

Fig. 3: Constant particle velocity curves.

If equation of Rayleigh line ($R = \rho_0^2 D^2 - (p_2 - p_0)/(V_0 - V_2) = 0$) is observed critically, first term is dependent on product of initial density and velocity, which is equivalent to mass flow rate through the shock wave. Since it is a square term, the value of mass flow rate is given by $\sqrt{[(p_2 - p_0)/(V_0 - V_2)]}$. Since mass flow rate is a real number, value within radical sign should not be less than zero. This indicates that if pressure rises while flow through a shock ($p_2 > p_0$), volume must reduce ($V_2 < V_0$). This situation depicts compression solution of detonation waves. Alternately, if pressure reduces while flow through a shock ($p_2 < p_0$), volume must increase ($V_2 > V_0$). This is expansion solution of deflagration waves.

Hugoniot Curves

For deriving expression for Hugoniot curves, equation for conservation of energy is also considered. 'E' represents specific internal energy and it depends on pressure and volume. In addition, it also depends on extent of completion of chemical reaction. This is shown by another term 'λ'. When this is 0, reaction has not started and when it is 1 reaction is complete.

$$E(p_0, V_0, \lambda = 0) + p_0V_0 + ½ D^2$$
$$= E(p_2, V_2, \lambda = 1) + p_2V_2 + ½ (D - u)^2$$

Or, $$E(p_2, V_2, \lambda = 1) - E(p_0, V_0, \lambda = 0)$$
$$= p_0V_0 + ½ D_2 - p_2V_2 - ½ (D - u)^2$$

[use mass conservation equation $\rho_0 D = \rho_2(D - u)$]

Or, $$E(p_2, V_2, \lambda = 1) - E(p_0, V_0, \lambda = 0)$$
$$= p_0V_0 - p_2V_2 + ½ D^2 - ½ (\rho_0/\rho_2)^2 D^2$$

Or, $$E(p_2, V_2, \lambda = 1) - E(p_0, V_0, \lambda = 0)$$
$$= p_0V_0 - p_2V_2 + ½ D^2 [1 - (\rho_0/\rho_2)^2]$$

[use equation of Rayleigh line $\rho_0^2 D^2 - (p_2 - p_0)/(V_0 - V_2) = 0$]

Or, $$E(p_2, V_2, \lambda = 1) - E(p_0, V_0, \lambda = 0)$$
$$= p_0V_0 - p_2V_2 + ½ [(p_2 - p_0)/(V_0 - V_2)] \times [1 - (\rho_0/\rho_2)^2]/\rho_0^2$$

Or, $$E(p_2, V_2, \lambda = 1) - E(p_0, V_0, \lambda = 0)$$
$$= p_0V_0 - p_2V_2 + ½ [(p_2 - p_0) \times (V_0 + V_2)]$$

Or, $$E(p_2, V_2, \lambda = 1) - E(p_0, V_0, \lambda = 0)$$
$$= p_0V_0 - p_2V_2 + ½ p_2V_0 + ½ p_2 V_2 - ½ p_0 V_0 - ½ p_0V_2$$

Or, $$E(p_2, V_2, \lambda = 1) - E(p_0, V_0, \lambda = 0)$$
$$= ½ p_0V_0 - ½ p_2V_2 + ½ p_2V_0 - ½ p_0V_2$$

Or, $$E(p_2, V_2, \lambda = 1) - E(p_0, V_0, \lambda = 0)$$
$$= ½ (p_0 + p_2) (V_0 - V_2)$$

Or $$E - E_0 = ½ (p_0 + p_2) (V_0 - V_2)$$

This is Hugoniot equation. For a given initial state, it represents all possible final state on the thermodynamic plane. Since all three conservation equations must be satisfied across the shock waves, the final state must lie on both Rayleigh line and Hugoniot line, if plotted on p–v plane.

The Hugoniot curve for the final state of product of detonation is shown by solid bold line in Fig. 4. The curve does not pass through the initial state. Rayleigh lines are also plotted on same scale. All Rayleigh lines are straight lines passing through point (p_0, V_0). As shock velocity increases, slope of Rayleigh line (although negative) increases. When shock velocity is low, there is no intersection of Hugoniot curve and Rayleigh line (for shock velocities D1 and D2). Under such circumstances, no solution is possible. For high shock velocity (say D4), two intersections are possible. One depicted by high pressure and high particle velocity (u2). This is called strong solution. Another is depicted by low velocity and low particle velocity and is called weak detonation. For certain critical shock velocity (in figure D3), Rayleigh line is tangent to Hugoniot curve. In this case only one detonation solution exists. This is called Chapman Jouguet velocity. Flow above this point is subsonic and below this is supersonic.

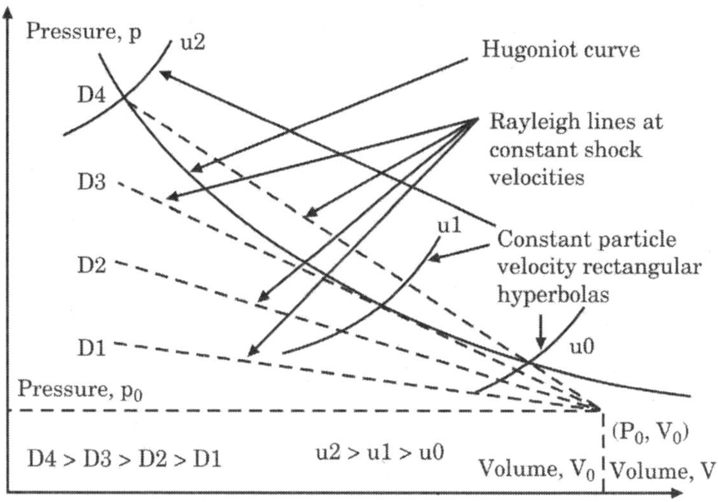

Fig. 4: Hugoniot and Rayleigh curves.

There are certain properties of Chapman-Jouguet (C – J) point.

- Rayleigh line is tangent to Hugoniot curve.
- $(dp/dV)_R = -(p - p_0)/(V_0 - V)$ and $(dp/dV)_H = 2\,(dE/dV)_H/(V_0 - V) = (p + p_0)/(V_0 - V)$. Equating both the slopes, $(dE/dV)_H = -p$.

- From thermodynamics, $dE = TdS - pdV$, or, $(dE/dV)_S = -p$. So isentropic lines are also tangent to both Hugoniot and Rayleigh line at C – J point.

Any propagation of disturbance through a compressible media can take place at a velocity equal to local sonic velocity or velocity of sound. The local velocity of sound can be derived from isentropic plot on p-v plane for the given compressible medium.

$$\text{Sonic velocity, } c^2 = -(1/\rho)^2 \times (dp/dv)_s$$

For an ideal detonation, the propagation velocity of small pressure disturbances (or sonic velocity) is equal to difference of detonation velocity and local particle velocity.

$$\text{Sonic velocity, } c = D - u_1 = D - (1 - \rho_0/\rho_1)D = (\rho_0/\rho_1) \times D$$

Combining above two expressions,

$$(dp/dv)_s = -\rho_0{}^2 D_2 = -(p_2 - p_0)/(V_2 - V_0)$$

This indicates that Rayleigh line is tangent to the isentropic-line of the reaction products at the C – J point. Additionally, for flow behind C – J plane, expansion of the reaction products takes place along the isentrope. In fact shock compression of explosives is a physical phenomena dominant event and chemical reaction does not get time to get executed during this process. So state of unreacted explosive at the detonation front is given by point of intersection of Rayleigh line and Hugoniot (shock adiabat) curves. This point is called Von-Neumann point and it occurs at a point at considerable higher pressure that C-J pressure. Shock adiabatic of unreacted explosives is a Hugoniot curve and at (v0,p0), slope of the curve indicates low pressure sound velocity in the unreacted explosive.

Further simplification is possible, if it is assumed that detonation products follow polytropic gas equation of state. Under these circumstances, specific internal energy (E) is equal to $pV/(\gamma - 1)$. Sonic velocity is also given by $\sqrt{\gamma pV}$. At C – J condition sonic velocity (c) is given by difference of detonation velocity (D) and particle velocity (u).

Sonic velocity, $c = D - u$

$$= (V_0)\sqrt{[(p_2 - p_0)/(V_0 - V_2)]} - \sqrt{[(p_2 - p_0) \times (V_0 - V_2)]}$$
$$= \sqrt{[(p_2 - p_0)/(V_0 - V_2)]} \times (V_0 - V_0 + V_2)$$
$$= (V_2)\sqrt{[(p_2 - p_0)/(V_0 - V_2)]} = \sqrt{(\gamma\, p_2 V_2)}$$

or, $V_2/(V_0 - V_2) = \gamma$

or, $V_0/(V_0 - V_2) = \gamma + 1$

or, $V_2/V_0 = \gamma/(\gamma + 1)$

Specific internal energy equation, $E - E_0 = \frac{1}{2}(p_0 + p_2)(V_0 - V_2)$

Or, $E_0 = E - \frac{1}{2}(p_0 + p_2)(V_0 - V_2)$

$$= p_2V_2/(\gamma - 1) - \tfrac{1}{2}(p_0 + p_2) \times V_0/(\gamma + 1)$$
$$= \gamma\, p_2V_0/(\gamma^2 - 1) - \tfrac{1}{2}(p_0 + p_2) \times V_0/(\gamma + 1)$$
$$= \gamma\, p_2V_0/(\gamma^2 - 1) - \tfrac{1}{2}p_2 \times V_0/(\gamma + 1)$$

[as $p_0 <<< p_2$]

$$= \tfrac{1}{2} \times p_2V_0/(\gamma - 1)$$

Now, $D_2^2 = (V_0)^2 \times [(p_2 - p_0)/(V_0 - V_2)]$

$$= (\gamma + 1) \times V_0 \times (p_2 - p_0) = (\gamma + 1) \times p_2 \times V_0$$

[as $p_0 <<< p_2$]

$$= 2(\gamma^2 - 1) \times E_0$$
$$= [(V_0)^2/(V_0 - V_2)^2] \times [(p_2 - p_0) \times (V_0 - V_2)]$$
$$= (\gamma + 1)^2 u_2^2$$

Mechanical Effects

High explosives are convenient and light source of energy to do mechanical work on the surrounding media. They generate high pressure, high temperature and high volume of detonation products, which act simultaneously on the surroundings effectively resulting in blasting effects. The power output from a typical explosive may be of the order of gigawatts per square centimeter (GW/cm^2). The instantaneous release of energy from explosives makes them an instant choice for rock blasting operations.

When a charge of explosive is detonated in air, it generates shock wave. This propagates through air at supersonic speed and is physically a discontinuous pressure front. The velocity of front may attain a velocity as high as 5-7 km/s at the point of detonation. As distance rises, both particle velocity and shock front velocity decreases. The detonation gas velocity follows particle velocity for certain distance (up to scaled distance of 1) from location of explosion and then it drops suddenly. After this point shock front velocity stabilizes, while particle velocity continues to fall in monotonic way. The variation of velocity with time can also be plotted, but it is customary to plot shock wave pressure with time. At each distance from the explosion site, peak pressures follow a decaying pattern and at a given location (constant distance from explosion site), pressure shows decaying feature. At given distance, at arrival time of shock wave, maximum pressure is recorded. As time progresses, pressure continue to fall till it reaches a value, less than atmospheric pressure. After a minimum pressure less than atmospheric pressure is attained, pressure rises again, attains atmospheric pressure at sufficiently long time step. In fact, for any explosion, sufficient time may be of the order of 10-20 milliseconds, only due to very fast detonation action.

When an axial stress wave propagates through cylindrical rods of steel, the amplitude of these waves reduces with time due to dispersion and internal friction in the steel. Suppose, a small charge is detonated in a cavity in an elastic solid, the rapid pressure in the solid generates shock wave. For a cylindrical charge, radial stresses at the wall remain constant with time. However, at larger radii in the material, radial stresses reduce with time. At sufficiently large radius, it may become less than zero also. The tangential stresses decrease with time and then stabilize at all radii. Tangential stress recorded at the surface of cylinder is lowest. The variation in axial stress is found independent of radii. In fact for rock blasting, this type of situation is generally encountered. Explosive is placed in a hole drilled in rock and exploded. Steel is relatively homogenous material and can be analyzed easily but behavior of rock under blast loading by explosion of explosive is an entirely new area. The rock blasting by explosive is practical application of explosive effect.

When explosive is filled entirely in the hole and is detonated, the walls are subjected to high radial acceleration and compression.

Although initial motion of the wall depends on the transient gas motion but later on, it becomes complicated due to other secondary effects and interactions. The transient instantaneous overpressure decays rapidly and actual deformation calculation starts with volume of pressurized gas in an undeformed chamber. The disappearance of shock wave (which occurs instantaneously) and work of expanding gases are to be separated. The equilibrium gas pressures may be of the order of 10-50 GPa and temperature is of the order of 3000-4000°C. So a shock front is generated in the surrounding media and pressure behind the shock decreases exponentially. Similar phenomena occur to shock waves propagation in air, lead-block and other materials, but isentropic expansion of detonation gas is utilized in a different way in those systems, after disappearance of shock wave.

If high density explosive is detonated in water, pressure rises rapidly due to compressibility. The water in the vicinity gets compressed and a shock wave with discontinuous front is generated. Initially the speed may be very high (above 4-5 km/s), but with increase in distance, it reduces to sonic velocity (1.5 km/s). The energy of shock wave will decrease and the thickness of shock wave shell increases with distance. After shockwave leaves, the gas bubble of detonation products continue to expand and impart radial kinetic energy to the surrounding water until pressure of the detonation gas drops below static pressure of the surrounding water. The radial velocity of water then decreases and under-pressure also prevails in the bubble. Water begins to move back towards the center thus giving rise to the pressure. So an oscillating gas sphere is obtained that eventually returns to initial volume. Mathematically, out of total chemical energy released by detonation of explosive, some part is taken away by the shock wave. During shock wave propagation, no compression of the medium takes place. Isentropic expansion of gas bubble from initial radius to some higher radius occurs and shock wave disappears. During this isentropic process, losses due to viscosity and heat loss may be neglected. During this period, volume change is conserved; meaning product of area and velocity of radial expansion of water is constant. Assuming a perfectly spherical gas bubble, $R^2 \times u$ = constant. Further calculation is started from this enhanced volume of the gas bubble after passage and passivation of shock waves.

In water, once shock waves are separated, the losses are small. The isentropic expansion work of detonation gas (pressurized) is utilized to impart kinetic energy to water around the bubble (confined gas mass). The prevalent water pressure prevents the radial expansion of water in outward direction. If depth of water is sufficient (high pressure), the bubble of detonation gas forms an oscillating system with slowly decreasing amplitude due to viscosity of water. If a low-density charge of same explosive is detonated in water, the shock wave energy decreases and more energy is available for doing mechanical work. Hence maximum kinetic energy is available, which increases with further decrease in density of explosive charge.

In a lead-block (standard test discussed elsewhere in the book for explosive strength assessment), expanding detonation gas imparts kinetic energy exactly in same manner as they do in water. However, lead is a solid and it absorbs some energy in deformation also. When energy of detonation gas is consumed in plastic deformation of hole in the lead-block, expansion of hole stops. The work done in this process on the surrounding atmosphere is negligible. Another situation is observed when explosive is partially filled in the hole or it is filled at lower density. Basically this gives some occluded gas in the cavity. The expansion of gas in this case may be isothermal phenomena initially, till it is distributed evenly in the expansion chamber. If initial cavity volume is increased keeping charge weight of the explosive same, (dissipating) shock wave energy becomes less intense. As total chemical energy from an explosive is constant, this means more energy is available for subsequent expansion work. So, volume of the lead block hole increases. This increase reaches a maximum value and stabilizes there. In the case of rocks, even if shock wave energy is fixed, critical burden at constant explosive quantity increases. It reaches a maximum. At decreasing intensity of shock wave due to increased cavity volume, the maximum is more pronounced.

In rock, porosity permits entry of gases in their pores. So in addition to deforming and compressing a bulk mass of material (rock in this case), the formation of crack and their propagation becomes another complicated event. No doubt gases in this case fail to impart any kinetic energy or velocity to the small particles, but detachment of rock from main mass may result in some velocity. The expansion of gases stops by creating some pores, cracks and main effect is loosening of rock particles and kinetic energy effect or enlargement of holes is not clearly visible.

Overall detonation is not a purely physical phenomena and augmentation of shock wave by chemical reaction becomes a prime criteria for such calculations. In nutshell, shock front evolved from pressure wave in a compressed material result in heating of explosives. The exothermic reaction takes some finite time, but even if it is assumed to have occurred in infinitesimal small time or instantly, the accompanying release of energy adds energy to the shock front. This drives shock front. However, the reactants (explosive), which after completion of chemical reaction become gaseous product. This gaseous product expands behind the shock wave, resulting in movement of a rarefaction wave into the shock front. The shock front, chemical reaction and the rarefaction wave remains in equilibrium and all move with same velocity called detonation velocity. So, there is no change in the shape of the shock front. At the same time detonation velocity also does not change with time and it remains constant. On a pressure-specific volume plot, Hugoniot-curves are drawn for the unreacted explosive and detonation products. From the point depicting initial condition of unreacted explosive on its Hugoniot, many Rayleigh lines can be drawn. For an ideal detonation given by C–J conditions, the Rayleigh line, which is tangent to Hugoniot of detonation product, is the real line and point of tangency is state of the detonation product behind the shock. Solution through pure theoretical considerations of physics alone may not be sufficient to solve and understand complete effects of detonation and shock waves from explosion of explosives or its mixtures.

3

Properties of Single Explosive Molecules

Explosives are compounds of carbon, hydrogen, nitrogen and oxygen, mainly. Occasionally, other ingredients like metals, fluorine, chlorine etc are also present, but there are a few explosive molecules of this variety. There are three main characteristics, which are needed for performance prediction of explosive. They are molecular formula, density and heat of formation. From molecular formula, molecular weight, oxygen balance etc are calculated. Density and heat of formation are bulk properties used for assessment of explosive formulations containing many compounds. Some of the popular explosives are given in Table 1. From the properties given in the table, other properties are derived. In the table, most of the properties are taken from various database. However, NTO is comparatively new entrant and its properties are taken from research paper from Lee *et al.*[20].

Calculation of molecular weight needs molecular formula of the molecules. Following molecular weights are taken for different elements.

Carbon	:	12.01
Hydrogen	:	1.008
Nitrogen	:	14.016
Oxygen	:	16.00

[20] Kien Yin Lee, Lonnie B. Chapman, Michael D. Coburn, "3-Nitro-1,2,4-triazol-5-One, A less Sensitive Explosive". *Journal of Energetic Materials,* Vol. 5, No. 1, 1987, pp. 27–33.

Table 1: Salient properties of some of the explosives

Explosive	*C*	*H*	*N*	*O*	*Density (g/cc)*	*Heat of formation (kcal/mol)*
RDX	3	6	6	6	1.806	14.71
TNT	7	5	3	6	1.654	–15.0
HMX	4	8	8	8	1.905	17.93
PETN	5	8	4	12	1.780	–128.7
Tetryl	7	5	5	8	1.730	4.67
HNS	14	6	6	12	1.740	18.7
NQ	1	4	4	2	1.810	–23.6
Explosive D	6	6	4	7	1.717	–94.0
TATB	6	6	6	6	1.938	–36.85
DATB	6	5	5	6	1.837	–23.6
NTO	2	2	4	3	1.930	–14.3

Molecular weight is directly calculated as weighed sum. Molecular weight of RDX = 3 × 12.01 + 6 × 1.008 + 6 × 14.016 + 6 × 16 = 222.174. Molecular weight of TNT = 7 × 12.01 + 5 × 1.008 + 3 × 14.016 + 6 × 16 = 227.158.

Oxygen balance is another parameter, which can be calculated for each of the given explosive molecule. A molecule is perfectly balanced, if it has sufficient oxygen to oxidize carbon-to carbon di-oxide and hydrogen to water. Generally numbers of oxygen molecules are either less or more and accordingly, oxygen balance is expressed as percentage of molecular weight of the molecule. For example, to calculate oxygen balance of RDX, its molecular formula is considered. It has 3 carbon atoms, which require 6 oxygen atoms for complete oxidation. It has 6 hydrogen atoms, which needs 3 oxygen atoms for complete oxidation to water. Total requirement of oxygen atoms is 9, but it has only 6 oxygen atoms available. There is a deficiency of 3 atoms, which has molecular weight of 3 × 16 = 48. For the molecular weight of 222.174, this deficient oxygen will give a negative oxygen balance of (– 48 × 100/ 222.174 =) –21.6%.

Oxygen balance of RDX is –21.6%.

TNT is given by molecular formula $C_7H_5N_3O_6$. Total oxygen requirement by carbon and hydrogen atoms is (7 × 2 + 5/2 =) 16.5, but it has only 6 oxygen atoms. This results in a deficiency of 10.5 atoms of oxygen. Oxygen balance is = –10.5 × 16/227.158 = – 4.62%.

Oxygen balance of TNT is – 4.62%.

For an explosive molecule with general formula $C_aH_bN_cO_d$, the oxygen balance is given by (d–2a–b/2) × 1600/MW. Oxygen balance is used along with density of explosive for the calculation of detonation velocity by empirical means. For such calculations definition of oxygen balance is slightly modified[21].

Modified oxygen balance = P = [(d–2a–b/2) ± w] × 100/n, where, n = number of atoms in the molecule, a, b, c are number of carbon, hydrogen and oxygen atoms in the molecule, w = sum of 'O'-atoms as per linkages; 0 for NO_2, C-Nitro, O-nitro, N-nitro, 1 for C-O-N in O-Nitro, 1.8 for C=O for carboxylic group, 2.2 for 'O' in phenols and alcohols. + sign if first part is positive, – sign if first term is negative.

Detonation velocity, D (m/s) = 2509 + 13.25 × P + 3793 × ρ + 12.81 × P × ρ, where ρ = density in g/cc.

Calculations are carried out for RDX for which a = 3, b = 6, d = 6, w = 0, n = 21 and ρ = 1.806 g/cc.

P = [(d–2a–b/2) ± w] × 100/n = –14.286

D = 8839.4 m/s against experimental value of 8850 m/s.

Calculation is repeated for TNT for which a = 7, b = 5, d = 6, w = 0, n = 21 and ρ = 1.654 g/cc.

P = [(d–2a–b/2) ± w] × 100/n = –50.0

[21] A.R. Martin, H.J. Yallop, "Some Aspects of Detonation, Part 1 – Detonation Velocity and Chemical Constitution". *Transaction of the Faraday Society*, Vol. 54, 1958, pp. 257–263.

D = 7060.7 m/s against experimental value of 6960 m/s.

Although values are matching to experimental values for these two molecules but this approach does not take into account the energy content of the explosive molecules and only oxygen balance may not be suitable indicator of detonation velocity. Attempts are made to include heat part also in the calculation of performance parameters of the explosives and details are presented in another chapter. Heat content of explosives is incorporated through a term called chemical heat release of the molecules.

The calculation of density and hest of formation for a given molecules are also discussed by various researchers. Qualitative structure-property relationship (QSPR) model is one method for such applications. Chemical information of the molecules is related to certain physicochemical properties of the molecules. This method depends on chemical structure and is independent of experimental properties. These models are used for calculation of density, heat of formation, enthalpy of formation and impact sensitivity.

In open literatures[22, 23], simple linear relation is developed for prediction of density. The main criterion is percentage weight of hydrogen in the molecule. It is found that high percentage of hydrogen in the molecule leads to lower density. *A linear relation is proposed between density and percentage hydrogen in the molecule, like* $\rho\,(g/cc) = A - B\,(\%H)$. The linearity is observed valid for hydrogen percentage between 0% and 6%. The percentage hydrogen is equal to number of hydrogen divided by molecular weight of the compound. The value of constants A and B depends on type of molecule considered. Structural groups govern value of these constants.

---

[22] L.T. Eremenko, "Interrelationship Between Density and Structure in an Explosive", Proceedings of the 11th Sumposium on Explosives and Pyrotechnics, Philadelphia, Pennsylvania, USA, Sept 15–17, 1981, pp. 29–1.

[23] L.T. Eremenko, "Density Calculation of Aromatic Crystals using Their Structure", Proceedings of the 8th International Pyrotechnics Seminar, Chicago, Illinois, July 12–18, 1982, pp. 740–758.

S.No.	*Nomenclature*	*A*	*B*
1.	Liquid aliphatic with relatively symmetrical substitution	1.780	0.096
2.	Liquid aliphatic with substituents not symmetrically placed around main chain	1.584	0.067
3.	Solid, noncyclic, aliphatic compounds containing only nitro and/or nitrate subsituents	2.114	0.169
4.	Solid noncyclic aliphatic secondary polynitroalkylamines and polynitroalkylamides	2.114	0.151
5.	Solid noncyclilc aliphatic secondary nitramines containing terminated ethylenenitrate and/or 2,2-dinitropropyl groups	2.114	0.134
6.	Solid noncyclic aliphatic primary nitramines and nitrates of primary amines	2.118	0.103
7.	Solid nitro and/or nitroxycyclanes and oxacyclanes	2.085	0.143
8.	Solid nitrazacyclanes and nitraazaoxacyclanes	2.086	0.093
9.	Solid normal (straight chain) nitrazaalkane	2.114	0.114
10.	Nonhydrogen substituents in an aromatic ring or rings of condensed or jointed aromatic systems	1.948	0.141
11.	Aromatic hydrocarbons, heteroaromatic compounds	1.954	0.130
12.	Aniline substituted on the ring or related heterocompounds having not more than one substituent or function on an amino nitrogen	1.984	0.124
13.	Compounds having substituents that form strong one-two dimensional intramolecular hydrogen bonds	2.094	0.132

From the above coefficients, select group in which a given compound falls and then using simple linear relation, density can be estimated. For example, RDX is a molecule of group 8. RDX has 6 hydrogen molecules. So, percentage hydrogen = 6 × 100/222.174 % = 2.7%.

Density of RDX in g/cc = 2.086 – 0.093 × 2.7 = 1.835 g/cc.

TNT lies in-group 11. It has 5 hydrogen molecules. So, percentage hydrogen = 5 × 100/227.158% = 2.201%.

Density of TNT in g/cc = 1.954 – 0.130 × 2.201 = 1.668 g/cc.

HNS ($C_{14}H_6N_6O_{12}$) belongs to group 10. Molecular weight of HNS = 12 × 14 + 1 × 6 + 14 × 6 + 16 × 12 = 450. So, percentage hydrogen = 6 × 100/450 = 1.33%.

Density of HNS = 1.948 – 0.141 × 1.333 = 1.76 g/cc.

The prediction is valid for explosives at 20°C and is within 2–3% accuracy. The limitations of this method are also enumerated in the paper. It is indicated that the method does not consider the state of matter and it is difficult to calculate density of explosives, whose melting point lies around 20°C. So, density of solid nitroglycerin cannot be found by this simple empirical relation. There are certain exclusions from the formulation also like primary N-nitroamines, acetal fluorinated in the carbon chain, ethers, polynitrocompounds with condensed aromatic rings, explosives containing difluoroamine and other explosophorous groups and chemical functions. Density of explosives with unknown combination of atoms cannot be estimated by this method.

For density of molecules, following relation is developed[24].

ρ (g/cc) = 1.75 RNDB + 0.018 × (HASA–2) – 0.70 RNCG + 0.054 × (HOMO–1) – 0.029 × KI3 + 1.99.

Where RNDB is relative number of double bonds, HASA-2 is the hydrogen acceptor surface area, RNCG is the relative negative charge, HOMO-1 is the energy of the second highest occupied molecular orbital, KI3 is the Kier shape index (order 3).

There are five descriptors in the model. RNDB is a constitutional parameter and is defined as ratio of number of double bonds to total number of bonds. This parameter gives detail about rigidity and hydrophobicity of the molecule. This parameter has the highest coefficient and is very important. Since double bonds can easily be compressed, higher value of this parameter leads to higher density as in case of aromatic explosives. HOMO-1 is a quantum chemical parameter and it reflects reactivity and polarizability of the molecule. Higher values of these parameters give larger density. RNCG is a function of chemical charge

[24] Feng Luan, Huitao Liu, Yingying Wen, Qingzhong Li, Xiaoyum Zhang, Jiazhong Sun, "QSPR Study for Estimation of Density of Some Aromatic Epxlosives by Multiple Linear Regression Approach". *Propellants Explosives Pyrotechnics*, Vol. 35, No. 2, 2010, pp. 169–174.

distribution in the molecule. It is mathematically expressed as ratio of maximum atomic partial negative surface charge to the sum of similar negative charges in the molecule. HASA-2 is a hydrogen acceptor surface area. RNCG and HASA-2 has coefficients with opposite signs depicting their reverse effects on density. KI3 is a topological parameter, which takes input from skeletal atoms, molecular branching and dispersion interaction of the molecule. Large branching and dispersion of molecules has negative influence on density and it leads to reduction in density, this is depicted correctly by negative coefficient in the formulation of HASA-2. For TATB, the calculation is reported to give density of 1.910 g/cc as against experimental density of 1.938 g/cc.

A high density is a top priority in seeking to develop new explosives. Nielsen[25] has listed some of molecular structural features that are known to increase density – high symmetry, quaternary and/or tertiary carbons, tertiary nitrogen, condensed rings. These requirements are met by polycyclic cage molecules such as tetrahedrane, triprismane, cubane, adamantine, wurtzitane, isowurtzitan etc and polyaza systems in which some carbons have been replaced by nitrogen. These molecules are strained and have high heat of formation, leading to greater heat release in the detonation reaction.

Chemical energy released (Q) from explosion of explosive molecule is another parameter of significant importance. It is given by following formula.

$$Q = \{[\Delta H_f]_{\text{explosive}} - [\Delta H_f]_{\text{products}}\}/\text{Molecular weight of explosive}$$

Most of the explosive-molecules contain carbon, hydrogen, nitrogen and oxygen. The calculation for any trial molecule given by chemical formula $C_aH_bN_cO_d$ needs discussion about product of explosion. The products of explosion of the class of explosives discussed are generally carbon, hydrogen, nitrogen, water, carbon monoxide, carbon dioxide and oxygen. Since pure elements have

[25] Nielsen A.T., "Polycyclic Amine Chemistry", Ch-5, p. 95–124, in Chemistry of Energetic Materials, *Edited* by Olah G.A., Squire D.R., Academics Press, San Diego, CA, 1991.

zero heat of formation, so heat of formation of the following substances only is important for calculation.

Carbon di-oxide	:	−393.5 kJ/mol gas
Carbon mono-oxide	:	−110.8 kJ/mol gas
Water	:	−242.0 kJ/mol gas
Water	:	−285.8 kJ/mol liquid

The physical state of product of combustion is also indicated above. Generally gaseous form of product of explosion is considered and water is considered in calculation of chemical energy release as water vapor. Two important aspects are to be illustrated here. First is conversion of heat unit. It is known that 1 kilocalorie (kcal) = 4.187 kilojoule (kJ). Second is conversion of calculation on mole basis to kilogram (kg) basis by using molecular weight of the materials. So heat of formation in alternate units for the substances can be calculated. For example, for carbon dioxide, molecular weight is 44. This means 1 mol of carbon dioxide will constitute 44 grams.

Heat of formation of carbon dioxide

$$= -393.5 \text{ kJ/mol} = -393.5/4.187 \text{ kcal/mol}$$
$$= -93.98 \text{ kcal/mol} = -93.98/44 \text{ kcal/g}$$
$$= -2.136 \text{ kcal/g}$$

Similar calculations can be made for other products of explosions. Molecular weight of carbon-mono-oxide and water may be taken as 28 and 18, respectively.

Heat of formation of carbon mono-oxide

$$= -110.8 \text{ kJ/mol} = -110.8/4.187 \text{ kcal/mol}$$
$$= -26.46 \text{ kcal/mol} = -26.46/28 \text{ kcal/g}$$
$$= -0.945 \text{ kcal/g}$$

Heat of formation of water

$$= -242.0 \text{ kJ/mol} = -242.0/4.187 \text{ kcal/mol}$$
$$= -57.79 \text{ kcal/mol} = -57.79/18 \text{ kcal/g}$$
$$= -3.210 \text{ kcal/g}$$

However, the prediction of heat of formation desires an estimation of product of explosion. Nitrogen is simply released as nitrogen gas, which do not contribute towards any heat of formation of products. Neither, it contributes for consumption of any oxygen. First of all, if enough oxygen is available, hydrogen is converted to water (H_2O). If explosive molecule is deficient in oxygen, hydrogen gas (H_2) will be released. If sufficient oxygen is there to convert all hydrogen to water, then remaining oxygen will try to oxidize carbon-to-carbon dioxide as per first scheme. As per another scheme, first carbon will be converted to carbon monooxide then in presence of excess oxygen carbon mono-oxide will be converted to carbon di-oxide.

Let us calculate chemical energy release by RDX. First scheme for estimation of product of explosion is utilized. The product of explosion will have 3 nitrogen molecules to consume all 6 nitrogen atoms. RDX has 6 hydrogen atoms, which will consume 3 oxygen atoms in formation of water. As it has 6 oxygen atoms, remaining 3 oxygen atoms will combine with 1.5 carbon atoms for the formation of carbon di-oxide. Remaining 1.5 carbon atom will remain unreacted as solid carbon in the reaction product. The chemical equation of this conversion is given below.

$$C_3H_6N_6O_6 \rightarrow 3N_2 + 3H_2O + 1.5CO_2 + 1.5C$$

Chemical energy release

$= (14.71 + 3 \times 57.79 + 1.5 \times 93.98)/222.174$ kcal/g

$= 1481$ cal/g: Multiply with density for change of unit

$= 1481 \times 1.805$ cal/cc $= 2673.3$ cal/cc

As per second scheme, formation of nitrogen and water will not be affected, but remain oxygen will try to oxidize carbon-to-carbon mono-oxide first. In this case, remaining 3 oxygen atoms, after oxidizing 6 hydrogen atoms, will oxidize carbon-to-carbon mono-oxide. No more oxygen is left for further conversion. The chemical reaction is shown below.

$$C_3H_6N_6O_6 \rightarrow 3N_2 + 3H_2O + 3CO$$

Chemical energy release
= (14.71 + 3 × 57.79 + 3 × 26.46)/222.174 kcal/g
= 1203.8 cal/g: Multiply with density for change of unit
= 1203.8 × 1.805 cal/cc = 2172.9 cal/cc

Another set of calculation is carried out for TNT.
First scheme:

$$C_7H_5N_3O_6 \rightarrow 1.5N_2 + 2.5H_2O + 1.75CO_2 + 5.25C$$

Chemical energy release
= (–15 + 2.5 × 57.79 + 1.75 × 93.98)/227.158 kcal/g
= 1293.98 cal/g: Change of unit
= 1293.98 × 1.654 cal/cc = 2140.2 cal/cc

Second scheme:

$$C_7H_5N_3O_6 \rightarrow 1.5N_2 + 2.5H_2O + 3.5CO + 3.5C$$

Chemical energy release
= (–15 + 2.5 × 57.79 + 3.5 × 26.46)/227.158 kcal/g
= 977.67 cal/g: Change of unit
= 977.67 × 1.654 cal/cc = 1617.06 cal/cc

Table 2: Calculated properties of explosive molecules

Explosives	*Molecular wt*	*OB (%)*	*Q Type I (cal/g)*	*Q Type I (cal/cc)*	*Q Type II (cal/g)*	*Q Type II (cal/cc)*
RDX	222.174	−21.6	1481.05	2674.77	1203.83	2174.12
TNT	227.158	−73.96	1293.99	2140.26	977.668	1617.06
HMX	296.232	−21.6	1475.36	2810.57	1198.15	2282.47
PETN	316.178	−10.12	1513.01	2693.16	1383.14	2462
Tetryl	287.19	−47.36	1419.23	2455.28	1026.06	1775.09
HNS	450.284	−67.51	1365.76	2376.42	955.419	1662.43
NQ	104.106	−30.74	883.523	1599.18	883.523	1599.18
Explosive D	246.172	−52	1085.95	1864.57	752.36	1291.8
TATB	258.204	−55.77	1074.69	2082.75	836.161	1620.48
DATB	243.18	−55.93	1173.37	2155.48	877.889	1612.68
NTO	130.1	−24.6	1056.65	2039.33	741.045	1430.22

For all the explosives given in Table 1, value of molecular weight, oxygen balance and chemical heat release by both the schemes discussed above is calculated and given in Table 2.

The calculations for each of the molecules follow the mathematical steps discussed above. The derived parameters are used to calculate detonation parameters and other performance parameters of explosive molecules. The properties are useful for estimation of performance of explosive formulations also.

However, these schemes are not universally accepted. Nitrogen anyway will be converted into nitrogen gas (N_2). As per another approach, first available oxygen is used to oxidize available carbon into carbon monoxide. If enough oxygen is not present, some solid carbon (C) remains. If all carbon is converted to carbon monoxide (CO), and oxygen is left, hydrogen starts converting into water (H_20). If hydrogen is more than available oxygen, hydrogen gas will evolve. If further after oxidizing all the hydrogen, some oxygen is left; carbon monoxide gets converted into carbon dioxide (CO_2). If TNT ($C_7H_5N_3O_6$) is decomposed as this approach, available oxygen (6) will oxidize carbon into carbon monoxide (6CO). Since number of carbon atom is more than that of oxygen, no oxygen is left after this conversion. Solid carbon (1C), Hydrogen gas (5/2 H_2) and nitrogen gas (3/2 N_2) will be other products of explosion.

$$C_7H_5N_3O_6 \rightarrow 1.5N_2 + 2.5H_2 + 6CO + 1C$$

Chemical energy release

$= (-15 + 6 \times 26.46)/227.158$ kcal/g

$= 632.86$ cal/g : Change of unit

$= 632.86 \times 1.654$ cal/cc $= 1046.75$ cal/cc

For polynitroaromatic compounds, heat of decomposition can be correlated to chemical energy release. Result of a regression analysis is given below[26].

---

[26] Zhang Housheng, Hu Rongzu, Yang Desuo, "Relationship between Heat of Decomposition and Heat of Explosion of Polynitroaromatic Compounds", Proceedings of the International Symposium on Pyrotechnics and Explosives". Beijing, China, Oct 12–15, 1987, pp. 448–453.

$Q_{dec} = 54.1 + 0.830 \times Q_{chem}$, where both heat parameters are in kJ/mol.

For Tetryl (Mol wt - 287.19), Chemical energy release is calculated in Table 2 as 1419.23 cal/g, which is equivalent to (1419.23 × 4.187 × 287.19/1000 =) 1706.57 kJ/mol. Heat of decomposition from the equation given above is 1470.5 kJ/mol.

Reported value of heat of decomposition is 1546 kJ/mol. The process is repeated for TNT (Mol wt - 227.158), for which chemical energy release is 1293.99 cal/g. The value is equivalent to 1230.72 kJ/mol. Calculated heat of decomposition from the formula is 1075.6 kJ/mol against reported value of 1074 kJ/mol.

The process is repeated for TATB (Mol wt - 258.204), for which chemical energy release is 1074.69 cal/g. The value is equivalent to 1161.8 kJ/mol. Calculated heat of decomposition from the formula is 1018.43 kJ/mol against reported value of 897 kJ/mol.

The process is repeated for HNS (Mol wt - 450.284), for which chemical energy release is 1365.76 cal/g. The value is equivalent to 2574.8 kJ/mol. Calculated heat of decomposition from the formula is 2191.2 kJ/mol against reported value of 1490 kJ/mol. The deviations are reported for compounds. This can be treated as approximate calculation for heat of decomposition of explosives.

Volume of gas produced per gram-mole of the explosives is also considered one of the important explosive parameter of explosives. The product of explosion becomes controlling parameter and it is observed that many factors affect this like initial density, temperature, degree of confinement, particle size, morphology, dimension of explosive charge, shape of the explosive charge, etc. Overall estimation of product of combustion becomes a dynamic problem and it becomes difficult to get actual composition of product of explosion. However, above two schemes for estimation of product of explosion can help to some extent. Since both the schemes result in different product of explosion, estimate by both the schemes may result in different volume of gases, as was seen for calculation of chemical energy. It is known that at standard condition, 1 mole of gas occupies a volume of 22.414 litres. Standard condition may be 1 atmospheric pressure (101325 kPa) and 25°C (298 K) temperature. Volume of product of detonation at standard condition may be obtained.

As per scheme II, RDX produces 9 moles of gas ($3N_2 + 3H_2O + 3CO$) and at standard condition, it has a volume of (22.414 litres × 9 =) 201.726 litres. So each gram of RDX (molecular weight = 222.174) will produce 907.964 cc (= 201.726 × 1000/222.174) of gas at standard pressure and temperature. However, as per scheme I, RDX produce only 7.5 moles ($3N_2 + 3H_2O + 1.5CO_2$) of gas with total volume of 168.1 litres (= 22.414 litres × 7.5). Each gram of RDX produces 756.64 cc of gas at standard pressure and temperature.

Table 3: Calculation of volume of gas evolved at STP

Explosives	*Scheme I*			*Scheme II*		
	No. of moles	*Volume (litres)*	*Vol/gram cc/gmol*	*No. of moles*	*Volume (litres)*	*Vol/gram cc/gmol*
RDX	7.5	168.1	756.64	9	201.7	907.96
TNT	5.75	128.9	567.36	7.5	168.1	740.03
HMX	10	224.1	756.64	12	268.9	907.96
PETN	10	224.1	708.90	11	246.5	779.79
Tetryl	7.75	173.7	604.85	10.5	235.3	819.48
HNS	10.5	235.3	522.66	15	336.2	746.66
NQ	4	89.6	861.20	4	89.6	861.20
Explosive D	7	156.9	637.35	9	201.7	819.45
TATB	7.5	168.1	651.05	9	201.7	781.27
DATB	6.75	151.3	622.15	8.5	190.5	783.45
NTO	4	89.6	689.13	5	112.1	861.41

Similarly for TNT, as per scheme I, number of moles of gases produced is 5.75 ($1.5N_2 + 2.5H_2O + 1.75CO_2$). So, volume of gases produced at STP (Standard temperature and pressure) is 128.88 liters and each gram of TNT will produce gas volume of 567.36 cc at STP. However as per scheme II, number of moles of product gases for TNT is 7.5 ($1.5N_2 + 2.5H_2O + 3.5CO$). Volume of gas produced at STP is 168.1 litres and each gram of TNT produces gas volume of 740.03 cc at STP.

Similar calculations can be repeated for all the molecules listed in Table 1. Calculations as per both the schemes give different values of gas volume (Table 3) at STP and for any further calculation; suitable and consistent scheme must be selected. Further calculations are carried out for scheme II only as it is more reliable for calculation of gas volume as well as for thermal estimates. In general, the product of explosion may include only four gaseous molecules namely N_2, H_2O, CO and CO_2. In very rare cases, H_2 or O_2 are obtained as product of explosion. For these gases, specific heat at constant pressure is dependent on temperature and a third order cubic equation describes specific heat at constant pressure for these molecules. In general, C_p (cal/gram-mole/K) = a + b × T + c × T^2. The values of coefficients of cubic equation for various possible products of detonation are given in Table 4. The table is obtained from tables of specific heat for these gases and is purely a matter of convenience.

Table 4: Coefficient of cubic polynomial for specific heats

Products	*a*	*b × 10³*	*c × 10⁶*
N_2	6.457	1.389	–0.069
H_2O	7.136	2.64	0.0459
CO	6.35	1.811	–0.2675
CO_2	6.339	10.14	–3.415
O_2	6.117	3.167	–1.005
H_2	6.946	-0.196	0.4757

From the table above, the value of specific heat at constant pressure for the product of explosion of each of the explosive molecule can be calculated as a cubic equation of temperature. Using scheme II, calculations are carried out for RDX. The governing equation is given below.

$$C_3H_6N_6O_6 \rightarrow 3N_2 + 3H_2O + 3CO$$

Total number of moles of product of explosion is 9. From the table given above,

C_p (cal/g-mol/K) = (3/9) × (6.457 + 1.389 × 10^{-3} T – 0.069 × 10^{-6} × T^2) + (3/9) × (7.136 + 2.64 × 10^{-3} T + 0.0459 × 10^{-6} × T^2) + (3/9) × (6.35 + 1.811 × 10^{-3} T – 0.2675 × 10^{-6} × T^2) = 6.648 + 0.00195T – 9.68667E – 08 T^2

Calculations are repeated for TNT. The equation is given below.

$$C_7H_5N_3O_6 \rightarrow 1.5N_2 + 2.5H_2O + 3.5CO + 3.5C$$

Total number of moles of product of explosion is 7.5. From the table given above,

C_p (cal/g-mol/K) = (1.5/7.5) × (6.457 + 1.389 × 10^{-3} T – 0.069 × 10^{-6} × T^2) + (2.5/7.5) × (7.136 + 2.64 × 10^{-3} T + 0.0459 × 10^{-6} × T^2) + (3.5/7.5) × (6.35 + 1.811 × 10^{-3} T – 0.2675 × 10^{-6} × T^2) = 6.6334 + 0.002003T – 1.23333E – 07 × T^2.

The chemical release of heat (Q in cal/g) from explosion of explosive is utilized for conversion of explosives into product of explosion and then rising temperature of gases to higher values. In other words Q = n × Cp × dT. Since Cp is expressed as cubic function of temperature, the expression can be written after integration within limits of standard atmospheric temperature (T_{298}) and flame temperature of product of explosion (T_f). Here it is worth noting that Q must be expressed in kcal/ gram-mole and the conversion of chemical heat release from one unit (cal/g) to other (kcal/g-mol) may be done by using molecular weight of the explosives.

Q (kcal/g-mole) = Q (cal/g) × Mol wt/1000

$Q = n \times [a (T_f - T_{298}) + (b/2) \times (T_f^2 - T_{298}^2) + (c/3) \times (T_f^3 - T^3_{298})$

Calculations are carried out for RDX.

Number of moles of gaseous products = 9

Molecular wt = 222.174

Q = 1203.83 cal/g = 1203.83 × 222.174/1000 kcal/g-mol

= 267.46 kcal/g-mol.

267.46 = 9 × [6.648 × (T_f – 298) + (0.00195/2) × (T_f^2 – 298^2) – (9.68667E – 08/3) × (T_f^3 – 298^3)]/1000.

This is a cubic equation, which can be solved by iteration or by close form solution. However, here an iterative approach is adopted.

For T_f = 3200 K, RHS of the equation = 253.03.

For T_f = 3500 K, RHS of the equation = 285.65.

$(T_f - 3200)/(3500 - 3200) = (267.46 - 253.03)/(285.65 - 253.03)$

T_f = 3332.71 K

For T_f = 3332.71 K, RHS of the equation = 267.33.

Further iteration is carried out and T_f is calculated as 3333.9 K.

The calculation is repeated for TNT also.

Number of moles of gaseous products = 7.5

Molecular wt = 227.158

Q = 977.668 cal/g = 977.668 × 227.158/1000 kcal/g-mol

= 222.09 kcal/g-mol.

$222.09 = 7.5 \times [6.6334 \times (T_f - 298) + (0.002003/2) \times (T_f^2 - 298^2) - (1.23333E - 07/3) \times (T_f^3 - 298^3)]/1000.$

This is a cubic equation, which can be solved by iteration or by close form solution. However, iterative approach is adopted.

For T_f = 3200 K, RHS of the equation = 210.53.

For T_f = 3500 K, RHS of the equation = 237.43.

$(T_f - 3200)/(3500 - 3200) = (222.09 - 210.53)/(237.43 - 210.53)$

T_f = 3328.9 K

For T_f = 3328.9 K, RHS of the equation = 221.99.

Further iteration is carried out and T_f is calculated as 3330 K.

The method described above can be utilized for the calculation of coefficients of cubic equation for specific heat at constant pressure for explosives and their mixtures and also for calculation of product temperature at constant pressure (Table 5).

Table 5: Calculation of detonation temperature

Explo-sives	*No. of moles*	*a*	*b*	*c*	*T*	*Q (Kcal/g mol)*
RDX	9	6.64767	0.001947	–9.687E-08	3334	267.46
TNT	7.5	6.6334	0.002003	–1.233E-07	3330	222.09
HMX	12	6.64767	0.001947	–9.687E-08	3322	354.93
PETN	11	6.65227	0.004307	–9.759E-07	4212	437.32
Tetryl	10.5	6.56262	0.001908	–1.456E-07	3282	294.67
HNS	15	6.5286	0.001892	–1.651E-07	3376	430.21
NQ	4	6.7965	0.002015	–1.155E-08	2658	91.98
Explosive D	9	6.63578	0.001994	–1.189E-07	2541	185.21
TATB	9	6.64767	0.001947	–9.687E-08	2846	215.9
DATB	8.5	6.61265	0.001931	–1.169E-07	2973	213.49
NTO	5	6.55	0.001808	–1.254E-07	2494	96.41

Further product temperature at constant volume can be calculated by multiplying previous calculated temperature by ratio of specific heats. For explosives, value of ratio of specific heats is generally taken as 3 and this can be utilized for calculation of product temperature at constant volume. Calculations are carried out for all the molecules given in Table 1 and the final results are given as Table 5. Scheme II is used for the calculations and the same is clear from the second column, which contains number of moles. The numerical values of coefficient 'a' and 'b' are positive, while the same for coefficient 'c' is negative for all the molecules considered. The product temperature at constant pressure is calculated, so that value of chemical heat release (Q) from the explosives matches with the heat absorbed by the products of detonation.

The calculation of production of detonation is still an arbitrary calculation based on certain assumptions of preferential oxidation of carbon and hydrogen. For calculation of product of explosion, water gas equation along with temperature dependent equilibrium constant is also employed. The water gas equation is given below.

$$H_2 + CO_2 \Leftrightarrow H_2O + CO$$

If concentration of each of the species is represented by keeping the chemical formula of the compounds in square brackets ([ ]), the equilibrium constant (K) for the equation is given below.

$$K = [CO][H_2O]/[H_2][CO_2]$$

This equilibrium constant is dependent on temperature and its variation with temperature can be represented by a quadratic equation as given below. The equation given below is valid for a temperature range of 1600 K to 5000 K. Explosion temperatures generally lies within this range for most of the explosives.

$$K = -0.00000063T^2 + 0.00601747T - 4.73751774$$

For any explosive based on carbon, hydrogen, nitrogen and oxygen, a general equation can be written as given below. Here values of 'a', 'b', 'c' and 'd' are known, as it is chemical formula of the considered explosive.

$$C_aH_bN_cO_d \rightarrow pCO_2 + qH_2O + rCO + sH_2 + tN_2$$

If above equation is balanced, mass conservation can be applied for each of the elements.

$$C : p + r = a$$

$$H : 2q + 2s = b$$

$$N : 2t = c \text{ or } t = c/2$$

$$O : 2p + q + r = d$$

An attempt is made to express, all the coefficients on right hand side of the equation in terms of 'p' and known coefficients.

So, $r = a - p$, $q = d - a - p$, $s = b/2 - d + a + p$.

Using water gas equation for the chemical decomposition of explosive,

$$K = [CO][H_2O]/[H_2][CO_2]$$

$$= rq/sp = (a - p)(d - a - p)/(b/2 - d + a + p)p$$

If certain temperature of explosion product is assumed, equilibrium constant (K) can be calculated. For a known temperature, above equation is quadratic equation in p.

$$(K - 1)p^2 + (bK/2 - dK + aK + d)p - a(d - a) = 0$$

Value of 'p' can be obtained from this quadratic equation. Once 'p' is known, 'q', 'r' and 's' can also be determined. So product of explosion at the assumed explosion temperature is known. From heat of formation and molar heat capacities of the products of explosion, explosion temperature can be calculated.

Assume temperature of explosion products of detonation of RDX as 3500 K. For RDX, Table 1,mentions, a = 3, b = 6, c = 6, d = 6, density = 1.805 g/cc, heat of formation = 14.71 kcal/mol. The molecular weight is specified in Table 2 as 222.174 g/g-mol. Equilibrium constant for the water gas equation can be calculated at the assumed temperature.

$$K = -0.00000063 \times 3500^2 + 0.00601747 \times 3500 - 4.73751774$$
$$= 8.60612726$$

The quadratic equation in terms of 'p' can be created and is given below.

$$7.606127p^2 + 6p - 9 = 0$$

So, $p = -6 + \sqrt{(6^2 + 4 \times 7.606127 \times 9)}/(2 \times 7.606127) = 0.76265$

Now, r = a – p = 3 – 0.76265 = 2.23735, q = d – a – p = 6 – 3 – 0.76265 = 2.23735, s = b/2 – d + a + p = 6/2 – 6 + 3 + 0.76265 = 0.76265, t = c/2 = 3.

The chemical equation for the explosion of RDX : $C_3H_6N_6O_6 \rightarrow 0.76265CO_2 + 2.23735H_2O + 2.23735CO + 0.76265H_2 + 3N_2$

After knowing trial product of explosion at the assumed temperature, the chemical heat release can be calculated.

Q = 1000 × [14.71 – 0.76265 × (–93.98) – 2.23735 × (–57.79) – 2.23735 × (–26.46)]/222.174 cal/g = 1237.231 cal/g.

This heat released during explosion is utilized in raising temperature of product of explosion to the assumed temperature.

As product of explosion changed, the value of specific heat at constant pressure also changes. So, heat needed by product of explosion to reach assumed temperature (3500 K) can be calculated from new specific heat at constant pressure.

$$C_p \text{ (cal/g-mol/K)} = 6.630634 + 0.002412T - 3.28706E - 07 \times T^2.$$

$$Q = 9 \times [6.630634 \times (3500 - 298) + (0.002412/2) \times (3500^2 - 298^2) - (3.28706E - 07/3) \times (3500^3 - 298^3)]/1000 = 1215.181 \text{ cal/g.}$$

The value of chemical heat release exceeds amount of heat needed to raise temperature of product of explosion to 3500 K. This indicates that assumed temperature is on lower side and higher temperature is to be assumed. The calculation can be repeated in same way as done earlier by interpolation between two assumed temperatures. After convergence, final value of explosion temperature for RDX is obtained as 3549.2 K. This value of explosion temperature is on higher side as compared to earlier calculation.

The process is repeated for TNT. Assume temperature of explosion products of detonation of TNT as 2000 K. For TNT, Table 1, mentions, a = 7, b = 5, c = 3, d = 6, density = 1.654 g/cc, heat of formation = –15 kcal/mol. The molecular weight is specified in Table 2 as 227.158 g/g-mol. Equilibrium constant for the water gas equation can be calculated at the assumed temperature.

$$K = -0.00000063 \times 2000^2 + 0.00601747 \times 2000 - 4.73751774$$
$$= 4.7774226$$

The quadratic equation in terms of ‘p’ can be created and is given below.

$$3.7774p^2 + 22.72098p + 7 = 0$$

So, $p = -0.32572.$

This negative value is not acceptable. So, the method fails. However, calculation is continued for calculation of explosion temperature.

Now, r = a – p = 7 + 0.32572 = 7.32572, q = d – a – p = 6 – 7 – 0.32572 = –0.67428, s = b/2 – d + a + p = 5/2 – 6 + 7 + 0.0.32572 = 0.3.174276, t = 3/2 = 1.5.

The chemical equation for the explosion of RDX : $C_3H_6N_6O_6 \rightarrow -0.32572CO_2 - 0.67428H_2O + 7.32572CO + 0.3174276H_2 + 1.5N_2$

After knowing trial product of explosion at the assumed temperature, the chemical heat release can be calculated.

$$Q = 1000 \times [-15 - (-0.32572) \times (-93.98) - (-0.67428) \times (-57.79) - 7.32572 \times (-26.46)]/227.158 \text{ cal/g} = 480.9899 \text{ cal/g}.$$

This heat released during explosion is utilized in raising temperature of product of explosion to the assumed temperature. As product of explosion changed, the value of specific heat at constant pressure also changes. So, heat needed by product of explosion to reach assumed temperature (2000 K) can be calculated from new specific heat at constant pressure.

$$C_p \text{ (cal/g-mol/K)} = 6.488725 + 0.000877T - 4.27726E - 08 \times T^2.$$

$$Q = 11 \times [6.488725 \times (2000 - 298) + (0.000877/2) \times (2000^2 - 298^2) - (4.27726E - 08/3) \times (2000^3 - 298^3)]/1000 = 599.6278 \text{ cal/g}.$$

The value of chemical heat release is lower than amount of heat needed to raise temperature of product of explosion to 2000 K. This indicates that assumed temperature is on higher side and lower temperature is to be assumed. The calculation can be repeated in same way as done earlier by interpolation between two assumed temperatures. After convergence, final value of explosion temperature for TNT is obtained as 1685.8 K. Since TNT has negative heat of formation and its oxygen balance is also on very negative side, the calculation may have resulted in stray temperature. However, it is clear from the calculation, that water gas equations are not very useful for calculation of explosion temperatures. The product of explosion calculated thereof may not be of much significance. Although water-gas equations are used for calculations related to gun propellants, but they have very little utility for the explosives.

In fact performance characteristics of explosives depend on method of evaluation and the end-use requirements. For shaped charge applications[27], high detonation velocity is expected, where as for rock blasting high explosive energy and expansion work (high volume of gessoes product of explosion) at low pressure is the requirement. Contrary to this in rocket propulsion, high energy and sonic velocity of the reaction products may play a major role. To understand behaviour of explosives, some parameters are developed by Andrew Block Bolten *et al.* [28] to correlate chemical and physical properties of explosives. The target property in the paper is constant volume explosion pressure. The developed factors are written as 'f' with sequential subscripts. Parallel to explanation, calculations are also shown for HMX ($C_4H_8N_8O_8$), which has a molecular weight of 296.232. Heat of formation is 17.93 kcal/mol.

First factor f_1 is related to heat or enthalpy of the molecule. More negative heat of formation (H_f), worse is the explosive and high negative heat of combustion (H_c), better is the explosive. These two parameters are accumulated together in factor f_1.

$$\text{Factor } f_1 = 1 - [(H_f - 1000)/(10H_c)]$$

Since this factor, uses combination of numerical value and variables (heat of formation and combustion), unit of variables becomes important for the numerical calculations. The paper emphasizes use of kJ/mol as preferred unit for calculation of factor. For HMX, H_f = 75.07 kJ/mol and H_c = −2626.3 kJ.mol. This gives value of f_1 as 0.9647816.

Second factor, f_2 can be called symmetry factor. It is based on the fact that if number of atoms of H, N and O are equal, then performance of explosive is on higher side. The value of this factor is cube of ratio of geometric mean to arithmetic mean of the number of hydrogen, nitrogen and oxygen atoms. Geometric mean is cube-

---

27 Himanshu Shekhar, Theoretical Modeling of Shaped Charges in Last Two Decades (1990-2010): A Review, *Central European Journal of Energetic Materials*, Vol. 9, No. 2, 2012, pp. 155–185.

28 Andrew Black Bolten, Jaimin Lee, "A Correlation between the Chemical and Physical Properties of C, H, N, O Explosives". *Journal of Energetic Materials* , Vol. 6, No. 1–2, 1988, pp. 107–128.

root of product of three numbers, while arithmetic mean is average of three numbers in this case. A more elaborate discussion on these is given in next chapter.

Factor f_2 = $[GM(H,O,N)/AM(H,O,N)]^3 = 27 \times (H \times N \times O)/(H + N + O)^3$

For HMX, number of nitrogen, hydrogen and oxygen atoms are all 8, so factor f_2 = 1. In fact, if number of hydrogen, oxygen and nitrogen atoms in a molecule are same, factor f_2 has the highest value and is 1. For all other cases, the value of factor f_2 is less than 1. For TNT, $C_7H_5N_3O_6$, the value of factor f_2 is $27 \times (5 \times 3 \times 6)/(5 + 3 + 6)3 = 0.88556$.

Third factor is a combined factor for three parameters. First part is ratio of nitrogen atom to total number of atoms. Second part is dependence on oxygen balance and third part is molecular weight of the material. Higher nitrogen content enhances explosive properties. Contrary to this, high molecular weight has negative influence on the material characteristics. For better performance, low negative or slightly positive oxygen balance is preferred. Since values of these three parts are to be correlated in such a way that each contributes in positive way to the end effects, factor f_3 has the following correlation.

Factor $f_3 = [100 \times 14.016 \times N/(C + H + N + O)] \times [1 - MW/1000]/(100 - OB)$

For HMX, value of this factor $f_3 = [100 \times 14.016 \times 8/(4 + 8 + 8 + 8)] \times [1 - 296.232/1000] / [100 - (-21.6)] = 400.457 \times 0.703768/121.6 = 2.31767$.

The fourth factor f_4 is a correlation for density and is given by square of density (ρ^2). For HMX $f_4 = 1.905^2 = 3.629$.

A combined factor, F is devised, which is log of product of all the factors.

$$F = \log (f_1 \times f_2 \times f_3 \times f_4) = \log (f_1 \times f_2 \times f_3) + 2 \log \rho$$

For HMX, value of $F = \log (0.9647816 \times 1 \times 2.31767 \times 3.629) = \log (8.11461) = 0.909267$.

It is indicated in the paper that explosive properties of different molecules are a function of this factor F. However, it is surprising that number of carbon atoms is not coming into picture anywhere in this correlation. Actually, explicit relations are not observed, but indirectly, number of carbon atoms contributes in heat of combustion (factor f_1) and also in oxygen balance (factor f_3). To stress utility of this factor F, constant volume explosion pressure for molecules is correlated using a linear equation and the value of coefficients are obtained by regression analysis. The same type of correlation is developed for V × MW/(1000T) also. The final equation after regression analysis is given below.

$$P(\text{atm}) = 23850 + 39493 \times \exp(1.674 \times F)$$

$$V \times MW/(1000 \times T) = 0.0222 + 0.031855 \times \exp(-1.242462 \times F)$$

For HMX, value of P (atm) = 23850 + 39493 × exp (1.674 × 0.909267) = 204803 atm and V × MW / (1000 × T) = 0.0222 + 0.031855 × exp (–1.242462 × 0.909267) = 0.0324929. It must be noted that parameter F is arbitrary and it simply expresses linearity. However, the values of various terms may be modified significantly. It is observed that explosive molecules with molecular weight lying between 200 and 300 can be represented well by the given linear equation. Errors are more with explosive molecules molecular weights outside this range. In calculation of f_4, sometimes density raised to power 2.5 or 3 gives better results.

Another utility of the developed correlation is estimation of density, if F is available for the given molecule. For, a molecule, if F = 0.97, it corresponds to constant volume explosion pressure (P) of 227427 atm. If first three factors f_1, f_2 and f_3 are same as that for HMX, the density of the molecule may be estimated as 2.0436 g/cc. This method of consolidating several parameters together is useful for prediction of explosive properties of molecules.

4

Properties of Explosive Formulation

Explosive Formulations

It is not explosive-molecules alone, but explosive compositions are also given some names. Cyclotol is the name given to RDX + TNT compositions. 75/25 cyclotol means the composition contains 75% by weight of RDX and 25% by weight of TNT. Composition B is desensitized 60/40 cyclotol with 1% by weight of wax. Overall composition B has 59.4% RDX by weight, 39.6% TNT and 1% wax. Similarly, octols are mixtures of HMX + TNT. Torpex is name given to aluminized cyclotols. Desensitized Torpex is called HBXs (high blast explosives). Minol is mixture of TNT + Al + AN. Pentolite is mixture of PETN + TNT. Picratol is a mixture of TNT + Explosive-D (picric acid). Tritonal is name of mixture of Al + TNT.

Some of the other TNT based explosive formulations, which are popular, are as follows:

Composition A : Pressed RDX + Wax

Composition A-3 : RDX/Wax 91/9

Composition A-5 : RDX/Stearic acid 99/1

Composition B : RDX/TNT/Wax, 59.4/39.6/1

Composition C : RDX + Binder

As far as applications are concerned, cyclotols or octols exhibit high detonation performance and brisance. They are used in metal accelerating applications like rockets and missiles warheads, shaped charges etc. HBX compositions can be optimized to show maximum underwater damage. They are also suitable for fragmentation and air-blast. Aluminized octols exhibits blast capabilities along with brisance. Picratol is insensitive in nature. All the mentioned compositions are used as main fillings. Compared to this, pentolite is suitable as booster compositions for initiation of main charges.

Single explosive molecule cannot give all the desired properties for a given situation. The requirement may be diverse. It may be high density, high velocity of detonation, ease of processing, ease of shaping, safety, directional properties etc. RDX is generally having very good detonation velocity and performance parameter but it has decomposition temperature and explosion temperature in the range of 204°C. It cannot be melted and shaped in proper form. Pressing it alone is a safety concern. Overall, it is always handled either in water or with some coating (of di-octyl phthalate or DOP) or in the form of a mixture with TNT. TNT can be obtained in molten state above 80°C, in which RDX can be safely added and then shaped in required form before solidification of TNT. Such mixtures of explosives are in more use than those of pure explosive molecules. It is necessary to understand method for calculation of explosive properties of such mixtures of explosives.

Properties of explosive formulations are dependent on their chemical constituents. Generally explosive formulations are described by percentage weight of their constituents. 75% RDX and 25% TNT composition means 100 g formulation will contain 75 g RDX and 25 g TNT. The percentage is generally specified in terms of weight and not volume. The calculation of chemical formula, oxygen balance, heat of formation, density etc of mixtures of explosive molecules can be derived using various mathematical means. Most of the time arithmetic or harmonic sums are used for calculation of properties. Based on this mathematical treatise on explosive formulations can be divided into two streams – (*i*) properties dependent on harmonic sum of individual explosive molecule properties and (*ii*) properties dependent on arithmetic sum of individual explosive molecules. Both are discussed in detail in the following sections.

Properties Calculated by Harmonic Mean

Harmonic mean is a number between two given number such that all three numbers form a Harmonic progression or series. Harmonic progression is a series, which gives an arithmetic progression, if each of its numbers is inverted. Arithmetic progression is a series of numbers, which maintain a common difference amongst any two successive numbers. For example, the series of numbers, 1, 4, 7, 10, 13, 16 .... are in arithmetic progression, because there exists a common difference of '3' between any two successive numbers. So, 1, ¼, 1/7, 1/10, 1/13, 1/16, ... form a Harmonic progression. If two numbers 'a' and 'b' are in harmonic progression and 'c' is Harmonic mean between the given numbers, then {a, c, b} forms a harmonic progression. 1/a, 1/c and 1/b form an arithmetic progression. This means 1/c is arithmetic mean (or average) of 1/a and 1/b. So, $1/c = [(1/a + 1/b) / 2]$ or $c = 2\,ab / (a+b)$. This is method to find harmonic mean between two numbers.

However, if more than two numbers not necessarily in harmonic progression exists, then harmonic mean can be calculated by taking weighted harmonic average. Let a,b,c,d,..... are numbers with weighed value of z,y,x,w,........ respectively and harmonic mean is h, then following relations exists for the numbers.

$$(z + y + x + w \ldots)/h = z/a + y/b + x/c + w/d + \ldots\ldots$$

or, $$h = (z + y + x + w \ldots)/(z/a + y/b + x/c + w/d + \ldots\ldots)$$

Calculation of theoretical maximum density (TMD) of the explosive formulation containing many ingredients is carried out using approach discussed above. Let us take a composition containing 75% RDX and 25% TNT. It is known that density of RDX and TNT are 1.806 g/cc and 1.654 g/cc respectively. Then

TMD of the formulation

$$= (75 + 25)/(75/1.806 + 25/1.654) \text{ g/cc}$$

$$= 1.765 \text{ g/cc}$$

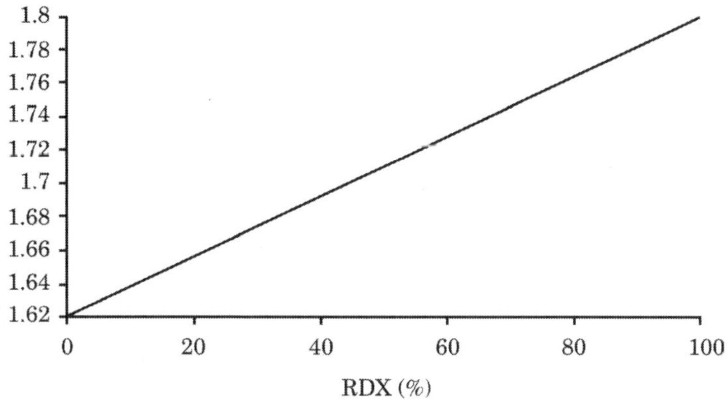

Fig. 5: Variation of density with % RDX in RDX/TNT formulations

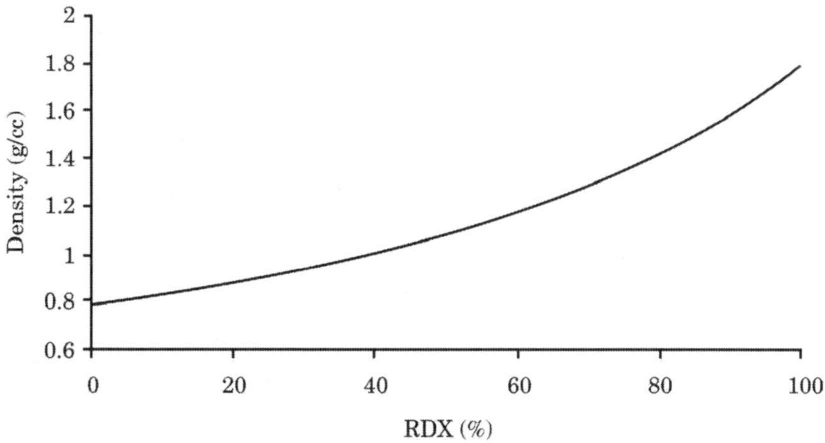

Fig. 6: Variation of density with %RDX in RDX/Wax formulation

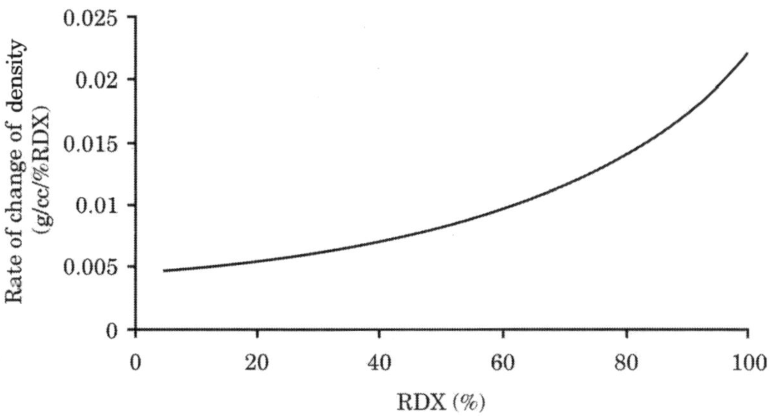

Fig. 7: Variation of rate of change of density with %RDX

For different percentage of RDX/TNT combinations a graph is plotted for variation of TMD with percentage RDX. The variation is shown in Fig. 5. Although the variation looks more or less linear but it is due to small difference between densities of RDX and TNT. As TNT has lower density, the density of explosive formulation rises with rise in percentage RDX. If materials with higher difference in densities are taken the non-linearity of the curve becomes obvious.

To show non-linear nature of variation of density with explosive formulation, RDX/wax formulations are considered. Density of wax is taken as 0.777 g/cc, which is much separated from the density of RDX (1.80 g/cc). Now variation of density is plotted for RDX/Wax formulations as Fig. 6. It is clear that variation is non-linear. At lower weight percentage of RDX (higher percentage of Wax), the formulation density is more dependent on wax density and rate of rise of density with RDX percentage is slower. However, at higher than 50% RDX content, the density rises at a faster pace. The rate of rise of density with percentage RDX for RDX/wax formulations is shown in Fig. 7. It is clear from Fig. 7, that density variation is faster at higher percentage of RDX. The above mathematical formulation is valid for binary system, but to get optimum properties, explosive-formulations contain more than two ingredients. The treatment for calculation of density follows similar steps.

Let us take an explosive formulation called TORPEX. This is an aluminized melt-cast explosive, used during World War II and it has following ingredients.

RDX (density = 1.80 g/cc): 42%

TNT (density = 1.62 g/cc): 40%

Al (density = 2.72 g/cc): 18%

The composition has ternary nature. However, mathematical calculation will not change due to more than two ingredients.

TMD of TORPEX is calculated as

$100/[42/1.80 + 40/1.62 + 18/2.72]$ g/cc

$= 1.83$ g/cc

Another explosive composition used during World War II is considered for calculation of density. Name of the composition is MINOL and it has following composition.

AN (density = 1.73 g/cc): 40%

TNT (density = 1.62 g/cc): 40%

Al (density = 2.72 g/cc): 20 %

The composition is also ternary in nature and mathematical calculations will be same as depicted for TORPEX.

TMD of MINOL is calculated as

100/[40/1.73 + 40/1.62 + 20/2.72] g/cc

= 1.812 g/cc

Another composition of World War II origin was H-6, which has four ingredients. The composition of the explosive formulation is given below.

RDX (density = 1.80 g/cc): 45%

TNT (density = 1.62 g/cc): 30%

Al (density = 2.72 g/cc): 20 %

Wax (density = 0.777 g/cc): 5%

TMD of H – 6 is calculated as 100/[45/1.80 + 30/1.62 + 20/2.72 + 5/0.777] g/cc = 1.745 g/cc.

Density of explosive charges can be calculated by making a cylinder of explosive formulation with known diameter and height. Volume of the charge can be calculated and weight can be taken on weighbridge of suitable sensitivity. Ratio of weight and volume gives practical measurement of density of explosive charges. But it is not always possible to make explosive charges in regular right circular cylinder form. Density of irregularly-shaped explosive blocks is also generally calculated by Archimedes principles using water displacement method. As per *Archimedes principles – "There is a reduction in weight of a substance on immersion in a fluid and reduction in weight is proportional to weight of fluid displaced".* For availability, cost-effectiveness, convenience and compatibility, fluid used is water for calculation/measurement of density. Most explosives formulations do not get dissolved in water. To utilize the principle of buoyancy proposed by Archimedes, weight of explosive piece (in grams) in air (W_1) and in water (W_2) is measured. Weight of explosive piece is equal to product of its density (ρ) and volume (V). As per Archimedes principles, $W_1 > W_2$ and reduction in weight on immersion in water is $W_1 - W_2$. Any material will displace water equal to its volume, so reduction in weight is equal to weight of that quantity of water, which is equal in volume to the volume of the explosive piece. In other words, reduction in weight is equal to product of density of water (1 g/cc) and volume of the test piece.

$W_1 = \rho \times V$ and $W_1 - W_2 = 1 \times V$.

Taking ratio, density of explosive, $\rho = W_1/(W_1 - W_2)$.

A more sophisticated method for density determination is by using Pycnometer. It has very high accuracy and 2-chamber system description is available in literature[29]. Pycnometer is an empty container generally, cylindrical in geometry. It's weight and volume (V) is known before testing any explosive. In this case, explosive is obtained in granular form by grinding and drying. Pycnometer tube is filled with explosive. The tube is evacuated to a level of 1 to 0.5 torr. The evacuated tube along with explosive-fill is weighed. Pycnometer weight is deducted and weight of explosive filled is obtained (M). Now pycnometer is filled with an operation liquid. The operational liquid may be n-alkanes (like heptane, hexane) or chlorinated hydrocarbons or saturated solution of measured substance in the solvent. The pycnometer is filled with solid explosive and operating liquid (in 1 : 3 ratio). The mixture is quickly frozen and evacuated. The content is melted and air is removed from the evacuated volume. The process is repeated till all micro-bubbles are removed from the pycnometer. Final weighment of pycnometer with operating liquid and solid explosive is obtained (N). If pycnometer liquid has a density of ρ, then

Density of the solid explosive $= M_\rho/(V_\rho - N)$

[29] P.L. Lagus, P.L. Turner, E.S. Gaffney, "A Rapid, Accurate Technique for Determining Densities of High explosives". *Propellants Explosives Pyrotechnics*, Vol. 2, No. 4, August 1977, pp. 81–84.

Another property of explosive formulation that follows Harmonic mean is molecular weight of the formulation. Molecular weigh of 75% RDX and 25% TNT formulation is calculated using the same mathematical analogy. Molecular weighs of RDX and TNT are 222 g/mol and 227 g/mol respectively. So, molecular weight of the formulation is 100/[75/222 + 25/227] = 223.23 g/mol.

Similarly for TORPEX, molecular weight of Al (MW = 27 g/mol) is needed additionally. Calculation follows similar steps and molecular weight of TORPEX is 100/[42/222 + 40/227 + 18/27] = 96.89 g/mol.

For MINOL, molecular weight of AN is taken as 80 g/mol. Molecular weight of the MINOL is 100/[40/80 + 40/227 + 20/27] = 70.57 g/mol.

For calculation of molecular weight of H-6, molecular weight of wax is taken as 14 g/mol. Molecular weight of the formulation is 100/[45/222 + 30/227 + 20/27 + 5/14] = 69.79 g/mol.

Properties Calculated by Arithmetic Mean

Although some of the explosive properties of ingredients contribute in the explosive formulations as Harmonic mean, but many other properties follow simple arithmetic mean, which is nothing but weighed average. Calculation of arithmetic mean for a given set of number follows following formulations. Let a,b,c,d,..... are numbers with weighed value of z,y,x,w,........ respectively and arithmetic mean is 's', then following relations exists for the numbers.

$$(z + y + x + w \ldots)*s = z*a + y*b + x*c + w*d + \ldots\ldots\ldots$$

$$\text{or, } s = (z*a + y*b + x*c + w*d + \ldots\ldots\ldots)/(z + y + x + w \ldots)$$

Same properties, which follow Harmonic mean route, cannot be treated with Arithmetic mean route. One property, which follows Arithmetic mean route, is heat of formation of explosive compositions. Heat of formation of explosive formulation can be calculated as weighed average of heat of formation of individual molecules. For calculation of heat of formation of a mixture of RDX and TNT, heat of formations of both is needed. The values for RDX and TNT are 14.71 kcal/mol and –15 kcal/mol respectively. Please note that the values are given in terms of per mol basis and it is apt to consider the values for a unit weight of the substances. Since RDX has a molecular weight of 222 g/mol, 75% of the weight will have 75/222 moles. Similarly for TNT number of moles will be 25/227. Now, heat of formation of 75% RDX and 25% TNT formulation is given by [(75/222) × 14.71 + (25/227) × (–15)] kcal/mol = 3.317 kcal/mol.

Calculation of equivalent formula for the mixture of explosive molecules also follows arithmetic mean type of variation. For the examples considered in previous section, relevant data are compiled as Table 6.

Table 6: Properties of explosive compositions

Name	*C*	*H*	*N*	*O*	*Mol wt*	*RDX/ TNT*	*Torpex*	*Minol*	*H-6*
Composition number						I	II	III	IV
RDX	3	6	6	6	222	75	42	0	45
TNT	7	5	3	6	227	25	40	40	30
AN	0	4	2	3	80	0	0	40	0
Wax	1	2	0	0	14	0	0	0	5
Al					27	0	18	20	20

For the composition number I, on 100 g basis, 75 g RDX is present, which is equivalent to 75/222 moles of RDX. 25 g TNT is equivalent to 25/227 moles of TNT.

Number of 'C' = (75/222) × 3 + (25/227) × 7 = 1.784

Number of 'H' = (75/222) × 6 + (25/227) × 5 = 2.578

Number of 'N' = (75/222) × 6 + (25/227) × 3 = 2.357

Number of 'O' = (75/222) × 6 + (25/227) × 6 = 2.688

So, equivalent formula of the composition is $C_{1.784}\,H_{2.578}\,N_{2.357}\,O_{2.688}$.

For the composition number II, on 100 g basis, 42 g RDX is present, which is equivalent to 42/222 moles of RDX. 40 g TNT is equivalent to 40/227 moles of TNT. 18 g of aluminium powder is equivalent to 18/27 moles of aluminium.

Number of 'C' = (42/222) × 3 + (40/227) × 7 = 1.801

Number of 'H' = (42/222) × 6 + (40/227) × 5 = 2.016

Number of 'N' = (42/222) × 6 + (40/227) × 3 = 1.664

Number of 'O' = (42/222) × 6 + (40/227) × 6 = 2.192

Number of 'Al' = (18/27) × 1 = 0.667

So, equivalent formula of the composition is $C_{1.801}\, H_{2.016}\, N_{1.664}\, O_{2.192}\, Al_{0.667}$.

For the composition number III, on 100 g basis, 40 g AN is present, which is equivalent to 40/80 moles of AN. 40 g TNT is equivalent to 40/227 moles of TNT. 20 g of aluminium powder is equivalent to 20/27 moles of aluminium.

Number of 'C' = (40/80) × 0 + (40/227) × 7 = 1.233

Number of 'H' = (40/80) × 4 + (40/227) × 5 = 2.881

Number of 'N' = (40/80) × 2 + (40/227) × 3 = 1.528

Number of 'O' = (40/80) × 3 + (40/227) × 6 = 2.557

Number of 'Al' = (20/27) × 1 = 0.741

So, equivalent formula of the composition is $C_{1.233}\, H_{2.881}\, N_{1.528}\, O_{2.557}\, Al_{0.741}$.

For the composition number IV, on 100 g basis, 45 g RDX is present, which is equivalent to 45/222 moles of AN. 30 g TNT is equivalent to 30/227 moles of TNT. 20 g of aluminium powder is equivalent to 20/27 moles of aluminium. 5 g of wax is equivalent to 5/14 moles of wax.

Number of 'C' = (45/222) × 3 + (30/227) × 7 + (5/14) × 1 = 1.890

Number of 'H' = (45/222) × 6 + (30/227) × 5 + (5/14) × 2 = 2.591

Number of 'N' = (45/222) × 6 + (40/227) × 3 = 1.745

Number of 'O' = (45/222) × 6 + (40/227) × 6 = 2.273

Number of 'Al' = (20/27) × 1 = 0.741

So, equivalent formula of the composition is $C_{1.890}\, H_{2.591}\, N_{1.745}\, O_{2.273}\, Al_{0.741}$.

All the molecular formula is calculated on mole basis assuming molecular weight of material as 100. However, for each of the composition molecular weight is already calculated in previous section. The number of molecules of each of the element will be multiplied by a factor given by molecular weight of the composition divided by 100. Molecular weight of the composition I is calculated as 223.23 g/mol. So, each number of molecules will be multiplied by 2.2323 in the equivalent formula to get the actual molecular formula of the composition. The equivalent molecular formula for composition I is given below.

Number of 'C' = 1.784 × 2.2323 = 3.9824

Number of 'H' = 2.578 × 2.2323 = 5.7548

Number of 'N' = 2.357 × 2.2323 = 5.2615

Number of 'O' = 2.688 × 2.2323 = 6.0000

So, equivalent formula of the composition is $C_{3.9824}\ H_{5.7548}\ N_{5.2615}\ O_{6.000}$.

Blast Wave from Explosives

Although single explosive molecules also produce blast waves, it is more apt to discuss it with explosive formulations. The more effective blast wave is produced, if metal powder is present in explosive formulations to augment shock and prolong its duration for enhanced effects at targets. Explosion refers to release of pressure waves of finite amplitude in air due to rapid release of energy. Even a source of initial finite pressure distribution may aggravate during flow through a compressible media till a discontinuity in pressure, density and temperature is practically realized. Such discontinuities are called shock fronts, which travels at supersonic speed. Simultaneous the particles in the media (air) are also accelerated due to passage of shock front, producing a net particle velocity in the direction of travel of shock. Compared to normal acoustic waves where

- Pressure change is infinitesimal
- Particles does not attain any velocity
- Disturbance moves at non-accelerating sonic velocity
- Linear process can define the motion
- Reflection and refraction follow normal rules

Shock wave propagation does not have any of the mentioned four characteristics. Sometimes these shock waves are also called blast waves.

If shock wave generated by any source, blast wave propagates in all direction away from the source in spherical fashion. The peak pressure of the wave diminishes with distance and also with respect to time. At any distance away from the source, blast wave takes certain finite time to reach. This time is called arrival time. Peak pressure is realized at arrival time on any given point. As time progresses, blast overpressure decays at a very fast pace to ambient pressure. This time duration during which medium is subjected to blast overpressure is called positive pressure phase. After this pressure goes below atmospheric pressure, which is long but have less magnitude. Impulse is sometimes calculated as area under the pressure-time curve at a point near the explosion site. There are several relations for estimation of such blast overpressure.

$P(t) = P_0 + P_p (1 - t/T)$, where, P_0 is atmospheric pressure, P_p is positive peak pressure, T = time duration of the positive phase, t is any instantaneous time.

This correlation assumed linear variation of pressure with time. This expression would result in underestimation of positive impulse. To improve this, exponential variation of pressure is also developed.

$P(t) = P_0 + P_p\, e^{-bt}$, where, P_0 is atmospheric pressure, P_p is positive peak pressure, b is a constant, t is any instantaneous time.

Sometimes combination of both the above results is also used for prediction of blast overpressure. This becomes a three parameter models with P_p, T and b as constants to be determined from experiments.

$$P(t) = P_0 + P_p (1 - t/T) \times e^{-bt/T}$$

During experimentation, it was observed that exponential decay term (c) decreases with time and some decrease rate of this exponential term is considered in a four-parameter model.

$$P(t) = P_0 + P_p (1 - t/T) \times e^{-b(1 - ct/T) \times t/T}$$

To get a better matching to experimental results, another version of four-parameter model is given below.

$$P(t) = P_0 + P_p (1 - t/T) \times e^{-b[1 + c/(1 + dt/T)]}$$

The most complex explosive blast equation is a 5-parameter equation given below.

$$P(t) = P_0 + P_p (1 - t/T) \times [ae^{-b\, t/T} + (1 - a)\, e^{-ct/T}]$$

Although negative pressure phase is of longer duration, it is not investigated exhaustively because of measurement difficulties. Computing characteristics in this phase is also slightly cumbersome. An empirical relation developed for negative pressure phase is as follows.

$$P(t) = P_0 + P_n (t/T) \times (1 - t/T) \times e^{-4t/T}$$

The pressure time profile is not the unique curve in explosion, density, temperature and particle velocity also follows similar trend with different amplitudes and duration. For the particle velocity of air by explosion of TNT, a complex empirical equation is developed.

$$u(t) = u_p (1 - at) \times e^{-bt} + c \times \ln(1 + \rho t)$$

In general experimental trials, any explosion may exhibit many-repeated shock of small amplitude after passage of main shock front. This may be due to successive implosion towards center of rarefaction waves from the contact surface between explosion products and air. Sometimes ground reflection may augment main shock front and relative higher peak overpressure is observed at longer distances.

Blast after colliding with a rigid plane wall, reflects and pressure in reflected waves is much higher than incident waves. In fact pressure, density and temperature all increases after reflection. If acoustic approximation is valid, the reflected overpressure is twice the incident overpressure. However, for strong incident shock wave, eight-fold enhancement in intensity is observed. A further higher limit of 20-fold rise is also possible in reflection of plane shock wave. The reflection of blast wave does not follow equality of incident and reflection angles. In addition to this, blast wave exhibits Mach reflection, which is different from conventional reflection principles. In a report[30] reflected as well as incident blast pressures are reported for RDX/Wax (95/5). The charge used was 50 g in 40 mm length and 30 mm diameter. The distance of measurement is kept 1 m to give scaled distance of 2.714 m/kg$^{1/3}$. The incident and reflected blast overpressure are 1.07 kg/cm^2 and 7.5 kg/cm^2 respectively. The results are average of minimum five trials. If scaled distance is reduced, higher amplification in pressure due to reflection is observed. In other words, if incident shock is stronger, amplification due to reflection is higher.

Blast wave scaling is generally expressed by Hopkinson or cube-root scaling law. As per this law, same blast overpressure is observed at a given distance from explosion site, if distance divided by cube root of energy (or mass) of the explosive is same. It indicates that explosion of 8 kg charge at distance of 2 m has same effect as 1 kg charge at 1 m and 27 kg charge at 3 m. This scaling law is useful in calculation of TNT equivalent, where explosion power of explosives is compared with that of TNT (heat of detonation = 1000 cal/g). This cube root scaling is applied to impulse and time also. By dimensional analysis, several groups are created so as to predict effect of explosion is unknown conditions from data at known conditions. Some of the nondimensional groups are as follows:

$R(p_a/E)^{1/3}$, p/p_a, $Ic/(Ep_0^{\,2})^{1/3}$, $Tc(p_0/E)^{1/3}$, where R = distance from explosion point, p_a = atmospheric pressure, E = energy of explosion (heat of detonation), p = instantaneous pressure, I = impulse from explosion, c = sonic velocity, t = time.

The quantities may be called scaled distance, scaled pressure, scaled impulse and scaled time respectively. To illustrate the utility of the equations, following example is cited. Suppose 0.512 kg TNT (heat of detonation 1000 cal/g) is detonated at normal atmospheric

[30] R.S. Punekar, Dr. B.M. Bohra, Dr. K.U.B. Rao, Report on Reflected Pressure Measurement Technique", Report No. HEMRL/ED/2003/2, Dated 18.07.2003, HEMRL, Pune, India.

condition with ambient pressure of 1 atmosphere and sonic velocity of 330 m/s. Peak overpressure and duration of positive phase at a distance of 3 m is recorded as 0.8 atmosphere and 1.8 milliseconds, respectively. The impulse of positive phase is 0.7 atm-milliseconds. If same explosive charge is exploded at an altitude where atmospheric pressure is 0.2 atmospheres and sonic velocity is 290 m/s, predict the performance.

Considering the data given for the explosive, two conditions are cited. For the first condition, E = weight × heat of detonation = 512 g × 1000 cal/g = 512 kcal, p_a = 1 atmosphere, c = 330 m/s, R = 3 m, p_s = 0.8 atmospheres, T = 1.8 ms, I = 0.7 atm-ms, Second condition has p_a = 0.2 atmosphere, c = 290 m/s.

$$\text{Scaled distance} = R(p_a/E)^{1/3} = 3(1/512)^{1/3}$$
$$= R\,(0.2/512)^{1/3}$$

So, $R = 5.13$ m.

This means that at high altitude where atmospheric pressure is low (0.2 atmospheres), same blast energy is generated at a longer distance (5.13 m) than that (3 m) for normal atmospheric condition.

$$\text{Scaled pressure} = p/p_a = 0.8/1 = p/0.2$$

So, $p = 0.16$ atmosphere.

This means same explosive will generate a pressure of 0.16 atmospheres only at high altitude, which was generating blast overpressures of 0.8 atmospheres at normal ambient pressures.

Scaled impulse = $Ic/(Ep_0^2)^{1/3}$
$= 0.7 \times 330/(512 \times 1^2)^{1/3} = I \times 290/(512 \times 0.2^2)^{1/3}$
So, I = 0.2724 atm-ms.

Scaled time = $Tc(p_0/E)^{1/3} = 1.8 \times 330 \times (1/512)^{1/3}$
$= T \times 290 \times (0.2/512)^{1/3}$
So, T = 3.5 ms.

Despite versatile universal uses, scaling law suffers from certain limitations.

- Rate dependent effects like viscosity of fluid and strain rates in solids are neglected.
- Gravitational effects are also neglected.

- Trajectory of missile generated by blast cannot be predicted accurately by scaling laws.
- Rise of fireballs or heated air cannot be predicted.
- The medium through which blast wave is transmitted must behave as ideal gas. For a strong shock, the law may be violated.
- The practical limits on small and large weights of explosives for experiments also restrict uses of these laws.

From explosive molecules point of view, addition of additives like metal powders (aluminium, zirconium, etc) may increase blast pressure from the explosive formulations. These inert materials absorb heat from the detonation products in reaction zone leading to formation of low heat of formation oxides. As oxygen gets consumed from the detonation products, adiabatic exponent and detonation pressure of resulting explosive formulation reduces. However, secondary reaction although may not contribute to the detonation wave but will strengthen shock wave in the surrounding medium to enhance blast effects. Practically, it is demonstrated that addition of aluminium and zirconium increases peak pressure by 10-15% and impulse by 20-30%[31]. The boron lowers both the quantities. Deflagration-to-detonation (DDT) transition behaviour of HMX in aluminized pressed explosive charges at 90% of theoretical maximum density (TMD) indicated that presence of aluminium slows down the DDT (deflagration to detonation transfer) process. Increase in percentage of aluminium results in enhancement in this effect also. Fine aluminium (5 micron) is more effective in this respect as compared to coarse powders (95 micron)[32]. Wax is found to have more inhibiting effect than aluminium. Such metalized explosives are useful for special situations where natural confinement is ensured as in underwater explosion or for demolishing air-targets at slightly longer distances.

[31] H.H. Licht, M. Frobose, "The Influence of Additives on Pressure and Momentum of the Air Shock from Detonating Charges". *Propellants Explosives Pyrotechnics*, Vol. 1, No. 1, April 1976, pp. 25–26.

[32] D. Price, A.R. Clairmont Jr, "Deflagration to Detonation Transition Behaviour of Aluminized HMX". *Propellants Explosives Pyrotechnics*, Vol. 4, No. 6, 1979, pp. 132–136.

5

Manufacture of Explosive Formulations

Raw Materials

Explosive formulations are manufactured by a variety of methods. The conventional explosive formulations based on TNT are generally manufactured by melt cast route. TNT has a melting point of 80°C and on heating to around 95°C, it can be converted to liquid form, in which solid explosive ingredients like RDX, HMX, PETN etc are added. After making a homogenous mix, the composition is cooled to get explosive mixtures. This process of manufacture is called melt casting of explosive. However, there are certain limitations on getting adequate density of melt-cast explosive formulations and most of the time 90 to 95% of TMD is only achieved. Compared to this, an alternate method by pressing of coated explosive ingredients in required forms leads to realization of pressed explosives. With development of technologies, plastic bonded explosives (PBX), which include polymeric binders to consolidate and keep together the explosive ingredients, are also developed. This is a cast-cure system and rather than physical change of state by temperature, chemical cross-linking is exploited in PBX formulations.

RDX or HMX is never handled in bare or uncoated form due to their high sensitivity. The sensitivity of these molecules depend on crystal size, crystal shape, crystal size distribution, crystal surface and crystal defects. Making explosive particles without corners or changing morphology to spherical form is one way to reduce sensitivity of explosives. Coating is another technique and wax, graphite, stearic acid and high polymers are used as coating agents for nitramines. But these ingredients are inert and tend to reduce the performance. Attempts are made to coat these molecules with insensitive explosives like 2,4,6-trinitrotoluene (TNT), 3-nitro-1,2,4-triazole-5-one (NTO) or 1,3,5-triamino-2,4,6-trinitrobenzene (TATB)[33]. HMX coated with composite of TNT and an energetic polymer (HP-1) by method of integrating solvent-nonsolvent with aqueous suspension melting gave heat of detonation comparable to bare sample and no deterioration is reported by such coating. Drop height and friction sensitivity are reported to be reduced by more than 50% by such coating[34].

In general, nitramines like RDX or HMX are either coated with DOP (Di-Octyl Phthalate) or they are transported in 40% weight ratio of TNT. Purpose of both is to reduce the sensitivity for ease and safety during handling. However, when these desensitized RDX or HMX is used in an explosive formulation, the desensitizer mass has to be reduced from explosive components. As these explosives are available in different particle sizes, it is natural to have different weights of coating in desensitized explosives. Coating thickness of DOP is generally taken as 1/30 times that of explosive mean radius. This indicates that bigger particle sizes of explosives are coated with higher thickness of coating materials. As a thumb rule, for DOP coated explosives (RDX or HMX) of any particle size, around 6% is taken as weight of DOP. This means every 100 g of coated RDX contains only 94 g of RDX and rest non-explosive ingredients. Such DOP coatings are preferred in sheet explosives and plastic bonded explosives formulations and proper compensation must be given to accommodate such loss in explosive quantity.

Suppose a nominal explosive formulation RDX/Binder 80/20 is processed in batch size of 1 kg. From the proportion given, RDX weight should be 800 grams and binder should be 200 grams.

---

[33] Chong-Wei An, Feng-Sheng Li, Xiao-Lan Song, Yi Wang, Xiao-De Guo, "Surface Coating of RDX with Composite of TNT and an Energetic-Polymer and its Safety Investigation". *Propellants Explosives Pyrotechnics*, Vol. 34, No. 5, 2009, pp. 400–405.

[34] Chong-Wei An, Jingyu Wang, Wenzheng Xu, Feng-Sheng Li, "Preparation and Properties of HMX coated with Composite of TNT/Energetic Material". *Propellants Explosives Pyrotechnics*, Vol. 35, No. 4, 2010, pp. 365–372.

However, in such formulation actual RDX quantity will be only 752 (= 8 × 94) grams. Performance of all explosive formulations is dependent on explosive content and DOP coating reduces effective explosive content. To compensate this, DOP coated RDX should be taken more than 800 grams so that effective quantity of RDX in 1 kg mix is 800 grams. Accordingly, quantity of RDX is calculated as 800 × 100/94 grams = 851.064 grams. If 200 grams of binder is used, then batch size will exceed 1 kg. Extra RDX used is 51.064 grams (= 851.064 – 800). Quantity of binder used in this formulation is accordingly reduced to 200 – 51.064 = 148.936 grams. So final quantity of ingredients in 1 kg mix is DOP coated RDX = 851.064 grams and binder = 148.936 grams. If both binder and plasticizers are in use, quantity of plasticizer is reduced by the excess quantity of DOP coated explosive ingredient.

Another sort of mathematical jugglery is carried out for melt-cast explosive formulations like RDX/TNT, TORPEX, or H-6, as elaborated in Table 6. In such cases, RDX is used, which is very sensitive. RDX cannot be transported or processed in bare form. One method to handle RDX safely is to store and transport it in water. However, extra water is to be carried, which has additional weight and volume. In addition to transportation inconvenience, water is to be removed from RDX at the time of processing. Most of the time a mix of RDX/TNT in ratio 80/20 is used as the basic raw material. The selection of such mix is purely for desensitization of RDX by coating it with TNT for ease of handling and transportation. TNT is used because it becomes part of final composition also. The ratio is fixed so that it is sufficient to coat RDX surface by TNT and on the other side, quantity of TNT should be minimum so that any other formulation can be processed by addition of TNT to this mix.

Since pure RDX is not available, all the explosive formulations have RDX / TNT 80 / 20 as basic mix. This mix may be named HS for ease of convenience in this context. The purpose of making this mix HS is as follows.

- Handling of pure sensitive explosive (RDX) is avoided.
- Formulations with less explosive percentage than 80% are possible to manufacture.

- This is least amount of TNT to be added as compromise between sensitivity and performance.
- Calculation of TNT for a given composition is easy from the given composition HS.

For preparation of x kg of RDX/TNT 75/25, it must contain 0.75x kg RDX and 0.25x kg of TNT. Since starting material is HS with TNT, assume that y kg of HS and z kg of TNT is used. Y kg of HS contains 0.80y kg RDX and 0.20y kg TNT. Balancing quantity of RDX, $0.75x = 0.80y$, or $x/0.80 = y/0.75$. Balancing quantity of TNT, $0.25x = 0.20y + z$, or $0.25x = (0.20 \times 0.75x / 0.80) + z$, or $x/0.80 = z / (0.25 \times 0.80 - 0.75 \times 0.20)$. Final equation is given below for calculation of y and z from known value of x.

$$x/0.80 = y/0.75 = z/(0.25 \times 0.80 - 0.75 \times 0.20).$$

As a general rule for a given composition, containing RDX/TNT in the ratio x/y, the governing equation is given below (HS is mixture of RDX/TNT in 80/20 ratio).

$$\text{(Final batch size)}/0.80 = \text{(Quantity of HS)}/x$$
$$= \text{(Quantity of TNT)}/(0.8y - 0.2x)$$

For preparation of 1 kg Torpex (refer Table 6 for composition), 180 grams of Aluminium is used. Quantity of RDX and TNT should be 420 grams and 400 grams respectively. However use of HS (80/20 mix of RDX/TNT) in place of RDX makes situation complicated, which is resolved by the derived equation indicated above.

Quantity of HS = $0.42 \times 1000/0.80$ grams = 525 grams

Quantity of TNT = $(0.8 \times 0.4 - 0.2 \times 0.42) \times 1000/0.80$ grams = 295 grams

There may be an alternate way for calculation of TNT, which is by method of difference. Since combined quantity of RDX and TNT in final formulation is 820 (420 + 400) grams, the total weight of HS and TNT should be 820 grams. Since HS is 525 grams, quantity of TNT is (820 – 525 =) 295 grams. This does not require complex calculation for TNT quantity.

To check that calculations lead to correct formulation, quantity of RDX and TNT is separately accounted. RDX is contributed by HS alone and out of 525 grams HS, quantity of RDX is 525 × 0.80 grams = 420 grams. Both HS and pure TNT contribute quantity of TNT. Quantity of TNT = 0.20 × 525 + 295 grams = 400 grams. The quantities of RDX and TNT are matching to the final requirements.

Similar calculations are repeated for composition H-6 (refer Table 6). 2 kg of H-6 contains 900 grams of RDX, 600 grams of TNT, 100 grams of wax and 400 grams of aluminium powder. For wax and aluminium powder, no further calculation is needed. Quantities of RDX and TNT are to be realized in terms of HS and TNT.

Quantity of HS = 0.45 × 2000/0.80 grams = 1125 grams.

Since total quantity of RDX and TNT is (900+600 =) 1500 grams, quantity of TNT is (1500 – 1125 =) 375 grams. To check accuracy of calculation step described above, formulation is used for calculation of quantity of TNT.

Quantity of TNT = (0.8 × 0.3 – 0.2 × 0.45) × 2000 / 0.80 grams = 375 grams

To check that calculations lead to correct formulation, quantity of RDX and TNT is separately accounted. RDX is contributed by HS alone and out of 1125 grams HS, quantity of RDX is 1125 × 0.80 grams = 900 grams. Both HS and pure TNT contribute quantity of TNT. Quantity of TNT = 0.20 × 1125 + 375 grams = 600 grams. The quantities of RDX and TNT are matching to the final requirements.

Casting of Explosive Formulations

After finalization of raw materials, processing of explosive formulations starts with melting of TNT at a temperature of around 90-95°C. Other ingredients are incorporated in the molten TNT at elevated temperature. The final composition is poured in molds and then it undergoes controlled cooling cycle to realize solidified explosive charge. The casting of molten explosive charge at elevated temperature gives melt-cast explosive formulations. Since only TNT remains in molten state, other ingredients do not affect processing parameters. After incorporation of all the ingredients, the explosive formulation is cast in tray and biscuits of explosive formulations are obtained on cooling. These biscuits are subjected to quality control for their density, percentage ingredient and after qualification only; they are remelted and cast in actual warheads, bombs, mines or other components. The density of molten TNT is around 1.44 g/cc, while solid TNT has a density of 1.654 g/cc. This leads to shrinkage by around 12-13% of explosive charge on cooling. The shrinkage is observed in the form of cavity at the center of the charge as piping. Since shrinkage cannot be avoided, some alternate measures are taken to avoid or reduce piping cavity. TNT is generally poured near its melting point in the mold. The temperature at the time of pouring in the molds is brought to 82-85°C. This makes TNT very viscous and chances of segregation and/or settling of high density ingredients are eliminated. However shrinkage on solidification cannot be avoided by this method. Sometimes solid biscuits of explosive formulations are directly poured in the molten mass of the explosive charges. This reduces shrinkage. The explosive filling is carried out in increments, so that first batch thickens before next installment is poured. Second filling already compensates for first filling shrinkage and so on. This reduces subsequent final shrinkage. Since solidification of explosive charges takes place from periphery towards center, the central portion contains voids or porosity. The molten mass is agitated by hot probes, so as to maintain uniform temperature and promote uniform cooling and avoid other casting defects. Hot probing also removes porosity at the center by releasing pressure head of filled mass. This leads to release of porosity faster. If provision for loading extra mass of molten explosive is available, more height of molten composition is filled than needed. On solidification height comes down due to shrinkage. Finally, filling topping mass of explosive composition after solidification of main mass compensates loss of mass due to shrinkage. Explosive charges by casting are also manufactured under vacuum casting and pressure casting.

The manufacturing of explosive composition always demands high density of the final charge for better performance. Sometimes pressure casting is also adopted for explosive manufacture. This is a combination of casting and pressing operations simultaneously. Introduction of vibration in pressure casting is another advancement suggested. In fact in addition to end requirement of high density, alternate methods are needed to cut short the time taken in normal melt-cast operation of explosive processing, where melting of TNT at elevated temperature, gradual cooling after casting followed by compensation for shrinkage of TNT charge may result in slow production rate of explosive charges. The application of pressure during casting changes whole mechanism of cooling from temperature reduction or heat extraction to that of raising of melting point by pressurization. Since hydrostatic pressure is same at all points during pressure casting, solidification starts at all the points in the entire molten volume. The solidification contour in the explosive charge are relative more homogeneous. In addition to this formation of voids are also controlled by plastic nature of TNT above 50°C. It is described that RDX/TNT 60/40 mixture is filled in molten condition in a preheated mold at 100°C. The pressure is raised to 1000 bar and temperature of the mixture was monitored. Application of pressure leads to immediate rise in temperature (~8°C) due to release of latent heat of fusion. The immediate solidification of molten mass is also confirmed by placing steel balls in the mix before pressurization. It is found that sinking of steel balls stopped at same moment at which pressure is applied. This indicates immediate solidification of molten mass on application of pressure. Void free charge and less variation in density is also observed experimentally[35].

In an attempt to understand the solidification of melt-cast explosive formulations, ultrasonic technique was employed for monitoring the online solidification process[36]. Such dynamic measurements can be useful for understanding complicated thermodynamics process of solidification, gaining knowledge on

---

[35] W. Witt, U. Sabranski, "A New Method for Cast-Loading Mixtures of Explosives". *Propellants Explosives Pyrotechnics*, Vol. 4, No. 1, 1979, p. 1–3.

[36] Weibin ZHANG, Yong TIAN, Hongping ZHOU, Zhanfeng YANG, Guan LUO, Hehou ZONG, Process Monitoring of Cast Explosive Solidification Molding by on-line Multichannel Ultrasonic Measurement, 17th World Conference on Nondestructive Testing, 25–28 Oct 2008, Shanghai, China.

the change rule of structure, performance and exploring mechanism of blowholes and cracks formation. Their ratio of TNT, RDX and additives in the mixture was 33.8:65:1.2. The dimension of mould was ϕ110 mm × 130 mm with wall 10 mm. The temperature of molding was 86°C and that of environment was 20°C. RDX is solid under 100°C. Data received from ultrasonic pulses indicated that solidification started from the bottom. The location of interface of melt-layer and solidified explosives was observed through the ultrasonic probe. Ultrasonic velocity and waveform were related to solidification velocity and physical characterization on the phase border. Ultrasonic attenuation was a function of the density and structure of the explosive that contained blowholes and cracks. The data indicate that ultrasonic transmission method can be used to reveal the regular of the solidification and monitor production's quality in real-time effectively.

Plastic bonded explosive (PBX) has a polymeric binder in place of melt-cast TNT and explosive molecules are dispersed in the binder. Both casting and pressing routes are used to manufacture PBX. For cast PBX formulations, binder and plasticizers are mixed in a vertical planetary mixer. Solid explosive particles are loaded in steps and mixing is continued. Mixing is continued at controlled temperature and sometimes under vacuum also. After mixing is complete, a curing agent is added and mixture is further mixed without vacuum. For HTPB-binder based PBX formulations containing RDX and HMX as explosive particulates, it is indicated that IDP (Isodecyl phelorganate) is a better plasticizer than DOA or DOS and higher solid loadings are possible by using IDP[37].

A more lucid description of PBX manufacturing is given in a paper, where different binders were tried for manufacture of cast PBX[38]. The formulation with energetic polymeric binder Fluorel

[37] M. Herrmann, Mikonsaari H. Krause, M. Kaiser, R. Huhn, M.H. Lefebvre, M. Alouaamari, C. Martin, A.E.D.M. can der Heijden, R.H.B. Bouma, LtKol Jacob Paap, S. Almada, "Particle Processing and Characterization: Part IV: PBX-Formulation and Characterization", 37th International Conference of ICT, Karlsruhe, Germany, June 27–30, 2006, p. 6–1

[38] Mostafa A. Radwan, "Sensitivity and Performance of Energetic Materials Based on Different Types of Energetic Binders", 37th International Conference of ICT, Karlsruhe, Germany, June 27–30, 2006, p. 56–1.

(hexafluoropropylene-vinyledene fluoride copolymer) is prepared by slurry technique. The polymeric binder is dissolved in solvent MEK (Methyl ethyl ketone) and lacquer is obtained. RDX is slowly added to the lacquer. Mixing is continued for 20-30 minutes. Distilled water is added drop-by-drop and stirring is continued for 30 minutes. Temperature is increased so that MEK evaporates. The formed beads are filtered, washed with water and dried at 70°C. HTPB and GAP based PBX formulations are prepared by cast technique. The pre-polymer and plasticizer is mixed together at 50–60°C. RDX is added and thoroughly mixed. At the end, HMDI is used as curing agent in the mixture. Slurry is poured in a mould and curing is continued for 2 weeks at 55°C.

Explosive Pressing

Pressing is another method for manufacture of explosives. This is used for manufacture of smaller charges. Some of the explosive formulations, mainly booster formulations are processed by pressing route. The composition contain RDX/wax in various proportions like 90/10, 95/5, 92.5/7.5, 97/3 etc. For processing, first the powder is loosely filled in the mold and then it is pressed to required density. During pressing, sufficient load as well as a dwell time of more than 20–30 seconds is applied. When loose powder is filled in the mold, the apparent density is low and length of powder column is more. When it is pressed density rises and column length of explosive reduces. After pressing, bottom base plate is removed and pressed pellet is ejected from bottom in a hollow spacer. High order of surface finish at the inner surface of mold and plunger is must for such processing else pressed explosive composition may break.

The main components of explosive pressing to get booster pellets are shown in Fig. 8–1. Since there is no machining on radius, the internal diameter of mold is matching to final dimension. Plunger dimension should be suitable for sliding fit in the mold body. Height of mold body should be made such that there must exist a gap of more than 20 mm at the top after filling loose explosive powder. The apparent density of explosive powder may be taken around 0.8 g/cc and height of mold body should be 20 to 25 mm more than height of loosely filled explosive powder. Spacer acts as arrester or plunger movement and movement of plunger once its flange touches spacer is physically restricted. Plunger length, mold height and spacer height are inter-related. Sum of plunger length and final pressed pellet height should be equal to sum of the mold height and spacer height. Base-plate is placed at the bottom to contain loose explosive initially as well as during pressing. Suitable mechanism to arrest lateral movement of base plate with respect to the mold body through step, protrusion, dowel-pin or other locating devices is needed.

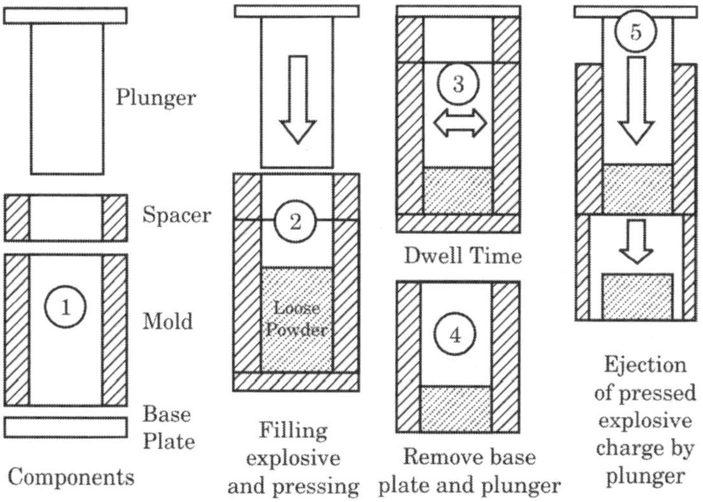

Fig. 8: Steps in pressing of explosive charges

First base plate, mold and spacer are assembled. Loose explosive powder is placed in the mold. Plunger is inserted and assembly is placed below ram of press (Fig. 8-2). Pressing is carried out and movement of piston is arrested once its flange touches spacer top. Plunger is left in pressed position for a dwell time of 20–30 seconds (Fig. 8-3). Then base-plate, plunger and spacer is removed (Fig. 8-4). A spacer with bigger cavity diameter and long length is placed below the mold with pressed pellet. Plunger is inserted again and pressed for ejection of pressed pellet (Fig. 8-5).

As an example, a trial case is considered, where final dimension of explosive pellet to be pressed is 10 mm diameter and 10 mm length with a density of around 1.70 g/cc. Internal diameter of the mold is taken 10 mm and mold thickness is taken around 20 mm. Although, thickness of mold can be calculated using thick cylinder formulation proposed by Lame's equation for internal pressures, but such calculations leads to very small thickness of the mold body, which is practically not feasible. By experience, mold body must be minimum 10-20 mm thick. Height of the loosely filled powder in the mold is calculated assuming apparent density of 0.8 g/cc.

Height of loosely filled powder = 10 mm × 1.75/0.8
= 21.25 mm.

So height of the mold is taken 25 mm extra as 45 mm. Mold dimensions are completely defined. Spacer internal and outer diameters are matching to that of the mold. Spacer height is considered as 25 mm. Spacer is completely defined now. Base plate thickness is taken as 25 mm. Height of plunger is calculated from heights of mold, pellet and spacer.

Height of the plunger = mold height + spacer height – pellet height = 45 + 25 – 10 mm = 60 mm.

Pressure levels to which pressing are generally permitted varies and there is no well set procedure to ascertain the same. As a thumb rule certain pressure levels may be specified as the highest pressure that can be applied on the loose explosive filling safely, which can result in required density. A pressure level of 1200–1500 kg/cm^2 is adequate for general explosive formulations. For the example considered, the maximum load applied using this thumb rule is (pressure × area = $1500 \times \pi \times 1^2/4$) = 1178 kg.

For manufacturing of small charges (L/D ~ 1), simple unidirectional pressing is sufficient and uniform density throughout the length of the charge is ensured. However, as length of the charge increases, explosives near plunger gets compressed more and the same pressing force is not transmitted to explosives located further away from the plunger. A general relation between pressure at different locations from applied pressure can be obtained.

Pressure, $P_x = P_a e^{-4fx/d}$, *where* P_a *= applied pressure,* f *= coefficient of friction,* x *= axial distance from point of application of the force,* d *= diameter pf the compacted cylinder of explosive material.*

The value of coefficient of friction may be 0.17 to 0.25. This results in density gradient along length in explosive charges. To overcome this, incremental pressing is suggested, where small lengths of explosive mass are compressed in succession. Bulk compaction of first increment in mould is carried out first. Then second installment of explosive granules is filled and pressing is carried out again. This second pressing results in further compaction of explosives filled during first increment. As pressure rises, density increases at a faster pace in the beginning and then it follows a linear curve with smaller slope. When unloading of first increment is carried out, density comes down in linear fashion. Further loading, during second increment compaction, density rises linearly again. During compaction, value of young's modulus, poisson's ratio or mechanical properties do not change significantly, but density variation is observed as per formula given below.

Change in density, $\Delta\rho = \rho \times \Delta P\,(1 - 2\nu)(1 + \nu)/[E(1 - \nu)]$, *where* E *= Young's modulus,* ν *= Poisson's ratio. The value of Poisson's ratio may be taken as 0.17.*

An advancement in unidirectional pressing of explosive charges is isostatic pressing. This is widely used as the safe method for compacting plastic bonded explosives due to non-rigid friction of the compacted explosive against any parts of the equipment. In this case, molding powder preheated to around 80°C is filled in a polyvinyl chloride (PVC) bag, which is fixed in an evacuating head and submerged in a fluid (liquid – either oil or water) medium. The fluid media is used for heating as well as for pressurization. First air entrapped in the explosive powder is evacuated down to pressure of around 2 torr. Then pressurization of PVC-bag through fluid medium is carried out in gradual manner. After keeping charge at the highest pressure for certain dwell time (~ 10 min), pressure is released gradually. The vacuum is released and then charges are recovered. Along with safety, this method is supposed to give highly compacted charges[39]. Sometimes temperature is also applied to augment compaction. From hydrostatic considerations, pressure-density relation follows following relationship.

Density, $\rho = \rho_r / [1 + (\rho_r / \rho_0 - 1) \times \{1 - (P/P_0)^{1/b}\}^{1/a}$, *where* ρ_r = *void free reference density,* ρ_0 = *initial density,* P_0 = *limiting pressure. Here value of 'a' may be around 0.5, while value of 'b' is around 3.*

Pressing of explosive charges desires preparation of moulding powder, if the booster pellets or any explosive charge is to be made of plastic bonded explosives (PBXs). In this case explosive molecules are coated with polymeric materials uniformly before final pressing to required densities. Manufacturing of moulding powder is a tricky affair as sensitivity of explosives and complete coating with polymers are to be addressed simultaneously. Explosives are always taken in water and polymeric material should be taken in a solvent. The regular agitation of mixture and simultaneous application of heat leads to coating of polymeric materials on the explosives. Invariably these solvents of polymers are flammable or toxic. A method for polymeric coating for manufacture of typical moulding powder is given in open literature[40]. The method described is polymer dispersion method, where explosive is taken in equal mass of water and is mechanically stirred before addition of polymeric material to this mixture. The polymeric material is an immiscible solution of the polymer in organic solvent. Sometimes plasticizer and thermal coagulation aid is also added. Then temperature is raised and stirring is continued. This completes thermal coagulation of polymers on explosives surfaces. The excess solvent is evaporated by raising

---

39 Tang Xingmin, Chen Yuanlu, Zhang Desasn, Zou Pinhuan, "A Simulation Study of the Explosion Accident in Isostatic Compacting HMX-PBX", Proceeding of the International Symposium on Pyrotechnics and Explosives, Beijing, China, Oct 12-15, 1987, pp. 337–342.

40 Ian J. Dagley, Luciana Montelli, Robert P. Parker, Veronica M. Siliva, "Pressed Booster Composition Prepared by Coating RDX using Polymer Dispersion". *Journal of Energetic Materials*, Vol. 10, No. 2–3, 1992, pp. 127–149.

temperature further and finally, the solution is cooled and then dried in the oven to give a freeflowing moulding powder. The moulding powder can be made free flowing by additional coating zinc stearate. This is achieved by adding hot aqueous solution of sodium stearate followed by a solution of zinc sulphate. Composition is washed thoroughly with water and then dried in vacuum at elevated temperature. In this process, special process equipments are needed for distillation. Coagulation technique using aqueous polyurethane, acrylic and ethyl-vinyl acetate dispersions are a viable alternative for the same.

RDX/Al/wax is also used extensively. However method of preparation and characterization HMX/Al/wax composition is elaborated[41]. The process of making moulding powder is described as desensitization of HMX, where wax is taken in excess hot water. After melting of wax (90°C), HMX is added in portion and stirring is continued. After complete addition and completion of mixing, the mixture is cooled and dried by suction, in an oven. The lumps are broken and material is sieved using 1 mm sieve. Mixing of Al-powder to this mix is also carried out at elevated temperature (90°C). Safety demands grounding of all equipments. Other safety precautions include stirring device powered by pressurized air, mixing of desensitized HMX with Al-powder in conductive polyethylene beaker, desensitization of HMX in water medium etc.

Moulding powder of PETN is investigated[42] and various crystal morphologies of pure crystals like flower and needle shapes in various particle sizes and surface areas are elaborated. Various polymers tried include ethyl cellulose, polyvinyl acetate, polystyrene, polymethyl methacrylate, vinyl chloride. The role of liquid medium is also elobarated as three fold in preparation of moulding powder – (*i*) To disperse and suspend explosive crystals (*ii*) To dissolve and dilute polymers and (*iii*) To act as Surfactants. The surfactants provide mechanism for the explosive molecules to attract polymers to despite on its surface rather than forming

41 P. Langen, P. Barth, "Investigation of the Explosive Properties of HMX/Al (70/30)". *Propellants Explosives Pyrotechnics*, Vol. 4, No. 6, 1979, pp. 129–131.

42 Marcia Dee Vannet, George L. Ball, "Pourability Enhancement of PETN Explosive Powders". *Journal of Energetic Materials*, Vol. 5, No. 1, 1987, pp. 35–55.

precipitated spheres of polymers alone. Purity, availability, cost, hygiene, purity, solubility are some of the issues, which goes in favour of deionized water as best liquid medium. The process of preparation of moulding powder is divided into three steps – (*i*) coating (*ii*) prilling or controlled agglomeration and (*iii*) stabilization and separation.

Explosive Sheet Manufacture

Another class of explosive charges is devised in the form of sheet explosives, which has a wide variety of applications in the field of welding, cladding and demolition devices. Sheet explosives are used for protection of tanks in explosive reactive armour (ERA), experimented exhaustively by Israel. The processing of explosives in sheet form needs an altogether different manufacturing technique. The mixing of ingredients, solvent treatment, rolling to required density, size and shape are realized in a highly skilled manner. Since solvent evaporation causes sever material defects like blowholes or entrapments and solvent vapours are health as well as fire hazards, generally solventless process is adopted for manufacture of sheet explosives. A typical manufacturing process for sheet explosive is explained in open literature[43]. All liquid ingredients including binder (HTPB), plasticizer (DOA), process aids (Lecithin), cure catalysts (FeAA) etc except curing agent (isocyanates) are added in a sigma blade mixer, where temperature during mixing is maintained at 30-40°C. Temperature is brought down to 25°C and curing agent is added. After mixing for duration of 30 minutes, the mixed dough is kept for partial curing in controlled humidity condition. The dough is rolled at ambient temperature to sheets of required thickness and is allowed to cure at room temperature for 2 days.

---

[43] Michael D. Joseph, Suresh K. Jangid, Ravindra S. Satpute, Bhagwan G. Polke, Tribhuran Nath, Shri N. Asthana, Alapati Subhananda Rao, "Studies on Advanced RDX/TATB Based Low Vulnerable Sheet Explosives with HTPB Binder". *Propellants Explosives Pyrotechnics*, Vol. 34, No. 4, 2009, pp. 326–330.

6

Initiation of Explosives

Most of the time explosive charges are initiated by electrical means with the help of a detonator. Although suitable initiation of explosives may be shock, impact, pressure and other mechanisms, but the major areas of initiation of high explosives of main charge invariably rely on explosive trains. In explosive train initial stimuli is given to initiatory charges. Initiatory charges are very sensitive and they are used in milligram-level in an explosive train. On initiation, it generates some shock pressure. The next component in the explosive train is booster, which enhances shock pressure manifold. They are slightly less sensitive and they cannot be initiated directly by low order stimuli as used to initiate initiatory. They are used in few grams level and are intermediate between main charge and initiator. Main charges contain secondary explosives of the order of few kilograms and they are relatively least sensitive. The enhanced shock energy by Booster is sufficient to initiate these explosives. From initiatory to main charge, sensitivity reduces and quantity increases. For successful initiation of main charge, each component of the explosive train should function effectively and they should produce sufficient shock at the main charge interface to make it detonate with maximum capacity. Although role of chemical reaction cannot be completely ruled out from detonation of explosives, the main phenomena is hydro-dynamic in nature with very little resemblance with combustion. The starting point may be burning, which occurs at an increasing rate for a few microseconds. A high pressure shock wave is rapidly formed, which sweeps explosive charge at high velocity. This leads to transfer of energy from reacted explosives to unreacted explosives by shock compression. So, rate of reaction is not dependent on normal heat transfer mechanism but it is dependent on shock wave transfer.

Explosives are classified as homogenous or liquid explosive and heterogeneous or solid explosive. Their methods of initiation by incident shock are quite different. Solid explosives are made of crystalline particles and air pockets. They have many internal surfaces in addition to well-defined external surface. When shock wave passes through a solid explosive, these random internal surfaces create hot spots or local temperature rise due to presence of air pockets. Liquid explosives are homogenous and they do not possess any internal secondary surface for creation of hot spot[44]. Liquid explosives behave as a continuum media for which mathematical formulations can be developed. In fact, if single crystal solid explosives are considered, there are no internal defects present and they behave in same way as liquid explosives. There are various differences between initiation of heterogeneous and homogenous explosive charges.

Imperfections or defects are less in homogenous materials (liquid explosives or single crystals). The reaction waves grow behind the shocked material and settles as stable detonation wave after overtaking shock. Heterogeneous explosives, made by pressing, casting, machining or extrusion contain fluid-mechanical irregularities, which act as origin of hot spots. These are associated with space-time fluctuations in the thermodynamics fields (pressure and temperature). In heterogeneous explosives, chemical reactions are triggered by shock and reaction waves originate from multi-points. So, the initiating shock grows in relatively smooth manner, which is in contrast to the abrupt changes that occur in homogeneous explosives.

Another parameter, which varies for these two classes of explosives, is diameter-effect, which is a curve between detonation velocity and inverse of charge diameter. For homogeneous explosives, decrease in detonation velocity near failure diameter

---

[44] R. Palmer, "Computer Calculations on the Initiation of High- Explosive Detonation". *The Computer Journal*, 6(1), 1963, pp. 39–43.

is small as compared to heterogeneous explosives. Another parameter is relation between input shock pressure and growth of detonation. For homogeneous explosives, small change in input pressure causes large change in time and distance of detonation. So, they are very sensitive to thermodynamic state of the explosives. Heterogeneous explosives initiates at much lower input pressure due to presence of multiple hot spot at all heterogeneity. The chemical energy release rate is less sensitive to state of heterogeneous explosives[45].

Sensitivity of Explosives

The initiation of explosives by different stimuli is characterized by sensitivity of explosives. A sensitive molecule gets initiated by less power or energy of stimuli, while reverse is true for an insensitive explosive. Sensitivity is an outcome of complex chemical process for an explosive molecule. It may be defined as ease of initiating a detectable chemical reaction in the molecule and its subsequent growth into high order detonation. The chemical kinetics and thermodynamics of molecules play major role in those calculations. Of course, various tests are carried out for sensitivity of different explosives, but impact height test is one of the most reliable tests.

For impact sensitivity, a given weight (preferably 2.5 kg) is dropped from different height on the sample of explosive (35 mg). The sample of explosive is placed on a polished flat tool steel anvil. The impact results in either an explosion (E) or a nonexplosion (NE). An explosion (E) at a given height ensures that for all heights above current height, explosion (E) occurs. A nonexplosion (NE) results in non-explosions (NE) for all the heights less than current height. For trial resulting in explosion (E), height is reduced in next trial, while height is increased, if non-explosion (NE) occurs.

---

[45] Stefan A Sheffield, Richard L. Gustavsen, Larry G. Hill, Ray Engelke, Robert R. Alcon, Lloyd L. Davis, "Insights into the Shock Initiation/Detonation of Homogeneous and Heterageneous HE", Invited paper for New Models and Hydrocodes for Shock Wave Processes in Condensed Matter, Paris France, 16–21 September 2001.

This is called Bruceton staircase method[46] of drop height variation. Minimum 25 samples are tested and height for 50% probability of explosion is estimated by statistical means. During testing by fall hammer apparatus, several variables play a vital role in results – like weight of fall hammer, quantity of explosive, hardness of anvil, morphology of explosive crystals. Force acting on explosive during fall-hammer test is numerically represented by a mathematical expression and it is found that force is a function of time and drop weight. Different force-strain laws are valid for different drop weights[47]. Height less than 20 cm may not give reliable results. It is a direct and simple method and is used to understand behaviour of explosive molecules under impact of moving mass. It simulates processing and handling needs also.

Kamlet and Adolph[48] have correlated 50% height of explosion of explosives with their oxygen balance. In fact oxygen balance is an indirect measure of energy available with the explosive. The formula for oxygen balance is given slightly modified and is reproduced below:

$$OB_{100} = 100 \times (2n_O - n_H - 2n_C - 2n_{COO})/MW$$

The sensitivity parameters are based on trigger linkage or privileged bonds concepts. In first set of molecules, C-NO_2 bond is given status of privileged bond and it breaks before alpha C-H bonds of polynitroaromatic rings. If alpha C-H bonds are not present, the following relation correlates impact height ($h_{50\%}$) to oxygen balance of the molecules. Several later researchers have modified the original relation and one of the relations is given below.

$$\text{Impact height, Log } h_{50\%}(\text{cm}) = 1.73 - 0.32 \times (OB_{100})$$

---

[46] Yin Menchao, Yu Tongchang, "Grouping Test – Up and Down Method of Variable Step for Sensitivity Tests", Proceedigns of the International Symposium on Pyrotechnics and Explosives, Beijing, China, Oct 12–15, 1987, pp. 847–852.

[47] R. Wild, "On the Time-Dependence of the Force in Tests of Impact Sensitivity". *Propellants Explosives Pyrotechnics*, Vol. 1, No. 3, August 1976, pp. 50–54.

[48] M.J. Kamlet, H.G. Adolph, "The Relationship of Impact Sensitivity with Structure of Organic High Explosives. Part II. Polynitroaromatic Explosives". *Propellants Explosives Pyrotechnics*, Vol. 4, No. 1, 1979, pp. 30–34.

It is observed that direct correlation is valid for similar molecules only. Diazophenols and aromatic azides are more sensitive to impact than polynitro aromatics and dependence of sensitivity on oxygen balance is also less pronounced. Similar is situation with nitramincs, where trigger linkages are $N\text{-}NO_2$, which imparts higher sensitivity to the molecule.

If alpha C-H linkages are presents, the coefficients of correlation change. Such compounds are more sensitive to impact than compounds without alpha C-H linkages. The new relation is given below.

Impact height, $\text{Log } h_{50\%}(\text{cm}) = 1.33 - 0.26 \times (OB_{100})$

For other classes of molecules also similar correlations are developed.

Nitroaliphatic: $\text{Log } h_{50\%}\,(\text{cm}) = 1.74 - 0.23 \times (OB_{100})$

Dintrofluoro: $\text{Log } h_{50\%}\,(\text{cm}) = 2.14 - 0.30 \times (OB_{100})$

Nitroamine: $\text{Log } h_{50\%}\,(\text{cm}) = 1.37 - 0.17 \times (OB_{100})$

For impact sensitivity, concept of activity index of molecular formula (F) is evolved and the formula for this parameter is given below[49].

$F = (100/MW) \times (4 \times n_{CNO2} + 4.75 \times n_{NNO2} + 5.75 \times n_{ONO2} + n_{OR} + 1.25\, n_E + 0.80 \times n_{AR} - 0.125 \times n_C - 0.0625 \times n_H)$, where MW = molecular weight of the molecule, n = number of groups, subscripts used are $C\text{-}NO_2$, $N\text{-}NO_2$, $O\text{-}NO_2$, (OR) carbonyl or C-O-R linkage: R - -H, $-NH_2$, - C etc, (E) ester bond acyl radical and acylaminde radical, (AR) Bezoid forms, Carbon, Hydrogen.

Impact height is correlated to this activity index (F) of the molecules by following relations.

Nitro : $\ln h_{50\%} = 14.6617 - 1.8904 \times F$

Nitramine : $\ln h_{50\%} = 11.4250 - 1.3820 \times F$

Nitrate ester : $\ln h_{50\%} = 11.6874 - 1.4154 \times F$

---

[49] Liu Dehui, Xing Yuming, "The Relationship Between Impact Sensitivity and Molecular Activity of High Explosives", Proceedings of the International Symposium on Pyrotechnics and Explosives, Beijing, China, Oct 12–15, 1987, pp. 800–807.

Keshavarz *et al.*[50] has developed another simple relation, where oxygen balance is not needed. It is observed that for C-HN-O explosives, the impact sensitivity can be directly correlated with number of atoms of carbon, hydrogen, nitrogen and oxygen. The relations are given below.

$$\text{Log } h_{50\%} = (X\,a + Y\,b + Z\,c + W\,d)/MW$$

Where a, b, c and d are number of carbon, hydrogen, nitrogen and oxygen atoms respectively. X, Y, Z and W are coefficients as per type of compounds and the values are given below:

For polynitroaromatics and benzofuroxans, X = 11.76, Y = 61.72, Z = 26.89, W = 11.48.

For polynitroaromatics with α-CH and α-N-CH and nitramines, X = 47.33, Y = 23.50, Z = 2.357, W = –1.105.

For polynitroaliphatics, X = 81.40, Y = 16.11, Z = –19.08, W = 1.089.

However, oxygen balance or number of respective atoms is not the only criteria for estimation of impact height. If it is so, then all isomers of a given molecules will have same impact height, which contradicts practical observations. Suppose tetranitronitro toluidine is considered. It has 4 nitro groups (NO_2), and one each of methyl (CH_3) and amine (NH_2) groups attached to the benzene ring. The location of methyl and amine groups gives ortho (adjacent), meta (one-off) and para (opposite) variants of the compounds. Their impact sensitivities are reported to be 36 cm, 37 cm and 47 cm respectively. Since in ortho and meta forms, three nitro groups are contiguous, they are more prone to chemical reaction by electronic and steric factors. Such pattern is not present in para form and it does not have such low impact sensitivity. As a rule, it can be stated that presence of three contiguous nitro groups in benzene ring may result in higher impact sensitivity. The three contiguous nitro group is not the only factor; interaction with

---

50 M.H. Keshavarz, H.R. Pouretedal, "Simple Empirical Method for Prediction of Impact Sensitivity of Selected Class of Explosives". *Journal of Hazardous Materials*, Vol. 124, No. 1–3, 30 Sept 2005, pp. 27–33.

adjacent methyl group also plays a major role. Consider three steric isomers of trinitrotoluene (TNT) – (*i*) 2,4,6 TNT (*ii*) 2,3,4 TNT (*iii*) 3,4,5 TNT. It is observed that second structure has three contiguous nitro groups and adjacent methyl group. It is the most sensitive with height of 50% explosion as 56 cm. The first molecule does not have three contiguous nitro groups but methyl group has two adjacent nitro groups. It has height of 50% explosion as 98 cm. However, the third molecule is least sensitive ($h_{50\%}$ = 107 cm) despite three contiguous nitro groups, probably because of no nitro group adjacent to methyl group.

Quantum chemical approach is also used for estimation of impact sensitivity of explosives. This utilized energy associated with molecular orbitals, ionization potential of atoms and total energy content of the molecules. The empirical relations are developed in one of the papers[51], using molecular mechanics at HF/STO-3G level of theory. A typical correlation is given below.

$\text{Log } h_{50\%} = 7.096 \times (E_{LUMO} - E_{HOMO} - MPP) + 0.000529 \times E + 0.711$, Where E_{LUMO} is energy of lowest unoccupied molecular orbital, E_{HOMO} is energy of highest occupied molecular orbital, MPP is mid-point potential of longest $C\text{-}NO_2$ bond (Weakest), E is total energy of the molecule.

For LUMO, orbital in lowest ground state of the molecule is considered that does not contain at least one electron. Lower E_{LUMO} means greater tendency of molecule to accept an electron and unstable molecule. This gives higher impact sensitivity. Contrary to this higher E_{HOMO} indicates capacity of molecule to easily donate an electron. Higher value indicates more reactivity. In fact lower value of E_{HOMO} means higher impact sensitivity. Difference of both the energies indicates faster propagation of reaction. More MPP gives higher impact sensitivity of the molecule and is correlated with $C\text{-}NO_2$ bond. In fact numerically, it is given by four times average of charges on C and N atoms divided by distance between these atoms (bond length). If total energy (E) is more, molecule will have more impact sensitivity.

---

51 N.R. Badder, C. Wei, A.A. Aldeeb, W.J. Rogers, M.S. Mannan, "Predicting the Impact Sensitivities of Polynitro Compounds using Quantum Chemical Descriptions". *Journal of Energetic Materials*, Vol. 24, No. 1, 2006, pp. 17–23.

For 1,3,5-Trinitrobenzene (TNB), E_{LUMO} = 0.12 eV, E_{HOMO} = –0.30 eV, MPP = 0.15 eV, E = –830.0 Hartree, Log $h_{50\%}$ = 2.1878 and $h_{50\%}$ = 154.1 cm. The experimentally observed value of Log $h_{50\%}$ is 2.000 for this molecule. The process is repeated for 2,4,6-Trinitrotoluene (TNT). For this molecule, E_{LUMO} = 0.12 eV, E_{HOMO} = –0.30 eV, MPP = 0.14 eV, E = –868.6 Hartree, Log $h_{50\%}$ = 2.238. The experimental value is found to be 2.196.

Further regression analysis is conducted and more correlations with elimination of certain independent variables are developed in the paper.

$$\text{Log}\, h_{50\%} = 7.485 \times E_{LUMO} - 5.922 \times MPP + 0.000423 \times E + 2.412$$

Using this relation, the value of Log $h_{50\%}$ for TNB and TNT are 2.0708 (Exptl = 2.000) and 2.1137 (Exptl = 2.196), respectively. These values are nearer to experimental values. Another correlation is also developed as given below.

$$\text{Log}\, h_{50\%} = 7.941 \times E_{LUMO} - 7.089 \times MPP + 2.085$$

Using this relation, the value of Log $h_{50\%}$ for TNB and TNT are 1.9745 (Exptl = 2.000) and 2.045 (Exptl = 2.196), respectively. These values are nearer to experimental values. For nitroaliphatic compounds, following correlation is proposed.

$$\text{Log}\, h_{50\%} = -9.757 \times MPP + 0.000922 \times E + 4.955$$

The correlation is further augmented with semiempirical PM3 with MOPAC. In those relations, ioniozation potential (IP) and dipole moments (DM) in the molecule are also added as independent variable. Ionization potential is the energy required to remove most loosely bonded electron from the outermost orbital of the atom in its lowest energy state. Actually IP is generally negative of E_{HOMO} of the molecule. High IP gives higher impact sensitivity. Dipole moment is a measure of asymmetry of molecular charge distribution. It is product of distance and charge quantity and it is directed towards positive charge. Impact sensitivity increases, if IP is high or if DM is high. Some of the developed correlations are given below.

$$h_{50\%} = 66.961 \times E_{HOMO} - 21.403 \times DM + 936.865$$
$$h_{50\%} = -66.961 \times IP - 21.403 \times DM + 936.954$$
$$\text{Log}\, h_{50\%} = -0.130 \times IP - 0.296 \times DM + 5.584$$

Under PM3 with MOPAC, the values of various independent variables are different from what is reported earlier. For TNB, $E_{HOMO} = -12.35$ eV, DM = 0.03 Debye, IP = 12.35 eV, the developed first two relations give $h_{50\%}$ as 109.254 cm, which is same as value of Log $h_{50\%}$ as 2.038 (Exptl = 2.000). For TNT, $E_{HOMO} = -11.92$ eV, DM = 1.38 Debye, IP = 11.92 eV, the developed first two relations give $h_{50\%}$ as 109.153 cm, which is same as value of Log $h_{50\%}$ as 2.038 (Exptl = 2.196). The last correlation gives value of Log $h_{50\%}$ for TNB and TNT as 3.968 (Exptl 2.000) and 3.626 (exptl 2.196) respectively. These values are not near the experimental results.

Another attempt is reported by W.P. Lai *et al.*[52], where in addition to conventional a, b, c and d terms for an explosive, two more terms are conceived for prediction of impact sensitivity. The relation is as follows:

$$\text{Log } h_{50\%} = [(X\,a + Y\,b + Z\,c + W\,d + F) / MW] + E$$

Where terms in first bracket [ ] have same meaning as for previous method. E is accommodation coefficient and F is sum of amending factor for each of the functional group interactions.

The coefficients have slightly different numerical values than earlier correlation.

For polynitroaromatics and benzofuraxans, X = 19.86, Y = 32.56, Z = –12.71, W = –13.32. E = 1.68. F is taken from table below.

Group	***Ortho***	***Meta***	***Para***
Nitro-nitro	–7.90	–20.28	–7.46
Nitro-amino	9.85	15.99	16.65
Nitro-hydroxyl	–2.64	–33.39	1.98
Amino-nitro	9.85	15.99	16.65
Amino-amino	—	14.72	—
Furaxan-amino		1.81	

[52] W.P. Lai, P. Lian, B.Z. Wang, Z.X. Ge, "New Correlations for Predicting Impact Sensitivities of Nitro Energetic Compounds". *Journal of Energetic Materials*, Vol. 28, No. 1, 2010, pp. 45–76.

For polynitroaromatics with α-CH and α-N-CH, X = 22.48, Y = 17.37, Z = −31.40, W = −48.87. E = 2.62. F is taken from table below.

Group	*Ortho*	*Meta*	*Para*
Nitro-nitro	-5.26	1.92	−11.36
Nitro-amino	15.31	15.96	15.37
Nitro-hydroxyl	15.26	—	30.53
Nitro-α-CH or α-NCH	−6.69	−0.69	−13.98
Amino-nitro	15.31	15.96	15.37

For polynitroaliphatic, X = 42.40, Y = 16.30, Z = −40.84, W = −31.53. E = 2.31. F is taken for interaction of N with different function groups from table below.

Group	*F*
$C\text{-}NO_2$	−6.15
$C\text{-}[(NO_2)_2]$	4.55
$C\text{-}[(NO_2)_3]$	−32.75
$[C\text{-}NO_2] - [C\text{-}NO_2]$	−89.14
$[C\text{-}NO_2]\text{ -C- }[C\text{-}NO_2]$	−56.50

For nitroamines or nitrate esters, X = 42.82, Y = 12.46, Z = −14.86, W = −21.47. E = 1.43. F is taken for interaction of N with different function groups from table below.

Group	*F*
$N\text{-}NO_2$ or $C\text{-}ONO_2$	4.52
$C\text{-}[(NO_2)_2]$	−4.20
$C\text{-}[(NO_2)_3]$	−38.03
$[C\text{-}NO_2]\text{ -C- }[N\text{-}NO_2]$	−28.19
$[N\text{-}NO_2]\text{ -C- }[N\text{-}NO_2]$	6.96
$[C\text{-}ONO_2] - [C\text{-}ONO_2]$	7.15
$[C\text{-}ONO_2]\text{ -C- }[C\text{-}ONO_2]$	−26.20
$[N\text{-}NO_2] - [C\text{-}ONO_2]$	−57.53

First of all, HNB (Hexanitrobenzene) is considered, which is poly-nitro-aromatic compound without any α-CH group. It is very sensitive and sensitivity is of the order of 11 cm. It has molecular formula $C_6N_6O_{12}$ and has molecular weight of 348.252. The calculations are carried out using Keshavarz method.

$$\text{Log } h_{50\%} = (11.76 \times 6 + 26.89 \times 6 + 11.48 \times 12)/348.252$$
$$= 1.0615$$

So, $h_{50\%} = 11.52$ cm

The method is repeated with method proposed by Lai *et al.* in this case, following special groups are present

meta Nitro-nitro groups – 3, coefficient = – 20.28
ortho Nitro-nitro group – 3, coefficient = – 7.90
para Nitro-nitro group – 3, coefficient = – 7.46
So, $F = 3 \times (-20.28) + 3 \times (-7.90) + 3 \times (-7.46) = -106.92$

$$\text{Log } h_{50\%} = [(19.86 \times 6 - 12.71 \times 6 - 13.32 \times 12 - 106.92)/348.252] + 1.68 = 1.0372$$

So, $h_{50\%} = 10.89$ cm.

Consider popular explosive molecule, 2,4,6-Trinitrotoluene (TNT) for calculation. It is has molecular formula of $C_7H_5N_3O_6$ and has molecular weight of 227.158. The experimental impact height is 98 cm for this. As this is polynitroromatic molecule with α-CH group, corresponding formula is used. Using Keshavarz method,

$$\text{Log } h_{50\%} = (47.33 \times 7 + 23.50 \times 5 + 2.357 \times 3 - 1.105 \times 6)/227.158$$
$$= 1.9777$$

So, $h_{50\%} = 95$ cm

The method is repeated with method proposed by Lai *et al.* in this case, following special groups are present

meta nitro-nitro groups – 3, coefficient = 1.92
ortho Nitro-a-CH group – 1, coefficient = –6.69
meta Nitro-a-CH group – 1, coefficient = –0.69
para Nitro-a-CH group – 1, coefficient = –13.98
So, $F = 3 \times 1.92 + 1 \times (-6.69) + 1 \times (-0.69) + 1 \times (-13.98) = -15.42$.

$$\text{Log } h_{50\%} = [(22.48 \times 7 + 17.37 \times 5 - 31.40 \times 3 - 48.87 \times 6 - 15.42)/227.158] + 2.62 = 1.92167$$

So, $h_{50\%} = 83.5$ cm.

The process is repeated for RDX for which experimental impact sensitivity is 26 cm. It is has molecular formula of $C_3H_6N_6O_6$ and has molecular weight of 222.174. It is a nitramine and the calculations can be repeated accordingly. Using Keshavarz method,

$$\text{Log } h_{50\%} = (47.33 \times 3 + 23.50 \times 6 + 2.357 \times 6 - 1.105 \times 6)/222.174 = 1.3075$$

So, $h_{50\%}$ = 20.3 cm

The method is repeated with method proposed by Lai *et al.* In this case, following special groups are present

$[N\text{-}NO_2]$ -C- $[N\text{-}NO_2]$ = 3, Coefficient = 6.96

F = 3 × 6.96 = 20.88

$$\text{Log } h_{50\%} = [(42.82 \times 3 + 12.46 \times 6 - 14.86 \times 6 - 21.47 \times 6 + 20.88)/222.174] + 1.43 = 1.4575$$

So, $h_{50\%}$ = 28.67 cm.

The impact sensitivity of explosives are correlated to other molecular properties also like electro-negativities, vibrational states, molecular masses, detonation gas concentration, oxidation number, partial atomic charges. Keshavarz[53] developed another very general correlation for nitro aromatics, benzofuraxans, nitroarmatics with alpha C-H, nitramines, nitroaliphatics, nitrate energetic molecules, based on structural parameters. The general correlation is given below.

$$\text{Log } h_{50\%} = (48.81a + 25.94\, b + 13.73\, c - 4.786\, d + 111.6\, DSSP - 132.3\, ISSP)/MW$$

Where a,b,c,d are number of carbon, hydrogen, nitrogen and oxygen atoms in the explosive molecule. DSSP and ISSP are decreasing and increasing sensitivity structural parameters.

DSSP = 0.7, 1.2, 1.7 for n_{NH2} = 1,2,3 in nitroaromatics and benzofuraxans

[53] M.H. Keshavarz, "Simple Relationship for Predicting Impact Sensitivitiy of Nitroaromatics, Nitramines and Nitroaliphatics". *Propellants Explosives Pyrotechnics*, Vol. 35, No. 2, 2010, pp. 175–181.

= 0.9 for –OH group [$n_{NH2} = n_{OH}$]
= 0.6 for benzofuraxans with –OH
= 0.7 if –OR attached to benzene ring
= 0.5 if $–O^-$ attached to benzene ring
= 1.2 if $–NH_2$ is also attached to benzene ring
= 1, if $n_{NO2} \geq 3$ in aromatic ring
= 0.8 for nitroaromatics with one $–CH_3$ and Phenyl at meta position
= 0.45 for nitramines containing = $C\text{-}N\text{-}NO_2$
= Number of –COO– in nitroaliphatics

ISSP = 0.5 if alpha C-H is present in nitroaromatics
= 0.7 for molecules with diazo or azido
= 0.5 for nitramines

In fact, final outcome of an explosive is always through ignition. Any method of initiation finally leads to ignition of explosive and the mechanism may be friction, pore collapse, adiabatic shear or other such complicated surface or bulk phenomena. The void collapse is preceded by formation of high speed jet, elevated velocities of convergent flow around cavity, shock heating by asymmetric collapse of cavity and compression of product in cavity. The hot spot theory is discussed in detail in next section. However, heated anvil drop hammer machine is reported to be used for testing thermal sensitivity of high explosives[54]. The reduction in drop height (molecules becomes more sensitive) with rise in temperature is experimentally demonstrated. For PETN drop height of 45 cm at 35°C reduces to 20 cm at 65°C. An additional feature of phase change can also be seen with impact height testing at different temperatures. It is reported that β-HMX is stable at room temperature and has the highest density of its 6 polymorphs. However, it undergoes a phase transition to a more sensitive δ-HMX at 435 K. This phase transition dependent impact height reduction can be observed with impact sensitivity testing machine with elevated temperature testing provisions.

---

[54] Gengxin Zhang, Brandon L. Weeks, "A Device for Testing Thermal Impact Sensitivity of High explosives". *Propellants Explosives Pyrotechnics*, Vol. 35, No. 5, 2010, pp. 440–445.

Structure – impact sensitivity correlation for different explosives were carried out through regression analysis and complex linear equation predicting dependence of impact sensitivity of nitro aromatic compounds on topology, number of nitro groups, amine groups and hetroatoms (like halogens)[55]. The topological term is found to have positive impact on impact sensitivity and as this term rises, impact height rises and compound becomes more and more stable. First effect is size and bigger aromatic molecules are stable. Conjugation also imparts stability and this is the reason why HNS, which is derived from TNT, is more stable than its precursor. Branching reduces topological parameter and reduces impact height. So if other parameters are same, sensitivity should increase with branching. Sensitivity increases with increase in number of nitro group and with reduction in number of amine group. Hetero-atoms have negative effects and adding any hetero-atom to explosive molecules, decreases impact height and makes compound more sensitive. The hetero-atoms may be active halogens (F, Cl) or oxygen. Although the regression equation is derived for trinitro benzene (TNB), the model based on perturbation molecular orbital theory within the Huckel molecular orbital approach is general and can be extended to other types of explosives.

Friction sensitivity is another sensitivity test, where different weights are rolled over sample placed on anvil. Friction is surface resistance to relative motion between to contacting bodies. It opposes motion and is equal to force needed to overcome adhesion and the interlocking of the asperities of two bodies. The frictional force is proportional to normal force, acting at the contacting surface and for each combination of surface textures, constant of proportionality varies. The energy supplied to overcome friction generally appears as heat.

The change in load is accomplished by changing weight and arm length of the instrument. The surface roughness also plays a major role in such experiments. Friction is usually present while mixing, poring, sieving, priming, consolidation, trimming, placing

---

[55] Lemi Turker, "Structure-Impact Sensitivity Relationship of Certain Explosive Compounds". *Journal of Energetic Materials*, Vol. 27, No. 2, 2009, pp. 94–109.

operations during processing, handling, storage or transportation of explosives. To simulate such conditions and assessment of safe operating procedure, several friction test apparatus and methods are developed in course of time. Some of them are given below[56].

- BAM friction test (Germany)
- Rotary friction test (Britain)
- ABL friction test (USA)
- Friction sensitivity test (Russia)
- Mallet friction test
- Torpedo friction test
- Emery friction test

Out of the given list, BAM friction test is very common. It was developed in German federal Republic in 1950s. Earlier pendulum devices were in use, where combination of rubbing and impact loading was observed. The BAM friction tester has a movable platform to which a roughened porcelain plate is attached. The plate is given oscillatory motion by an electrical motor. A fixed porcelain peg (mallet, hammer) rubs on this plate. Peg is connected to a lever of third type with fulcrum at one end and applied load at the other. The contact between peg and plate occurs in between. By placing different loads at free end of the lever, different loads are applied on contact points. The explosives are placed on the porcelain plate at the position where peg rubs against plate. Testing for different explosives indicated that as crystal density of primary explosives increases, friction sensitivity level comes down[57].

Sample, placed on porcelain table is subjected to friction twice by the reciprocating motion of another porcelain peg loaded with certain weight. The surface roughness of porcelain plays a major

---

[56] Glen Miles, "The development of an Elevated Temperature Friction Sensitiveness Test",

[57] Joel Harris, "The Friction Sensitivity of Primary Explosives", Proceedings of 11th Symposium on Explosives and Pyrotechnics, Philadelphia, Pennsylvania, USA, Sept 15–17, 1981, p. 6–1.

role in numerical value of friction sensitivity[58]. Use of emery paper of constant surface roughness is advocated in the paper. It is observed that pure azides are not very sensitive to slight variation in roughness of rubbing surfaces; however, better consistency in values is reported for friction sensitivity testing with emery paper. In USA, friction pendulum is used in earlier days, where steel or fibers are used to rub the explosive samples. The reaction behaviour is classified as explosion, snaps, cracks or unaffected. The second method was roto-friction test apparatus. It uses a flat steel rod on the surface of an energetic material using a torque converter mechanism. The sample (20 mg) is placed on sample holder and spinning rod is placed on the sample. Time is recorded from the time the rod contacts the sample until ignition. If no ignition occurs, time till rod is lifted is accounted. The energy value is calculated when visible or audible ignition occurs. The energy imparted to explosive sample is calculated by formula given below.

Energy, $E = \pi W w r t$, where, W = angular velocity, w = weight producing torque, r = radius at which weight is applied, t = time of explosion.

For various pyrotechnic compositions, a general relation in the form of following equation is developed from experimental data.

Rate of heating, $q = NW/(aN + b)$, where, N = load on friction rod, W = angular velocity, a = infinite rod coefficient, b = zero load coefficient.

The formulation is found applicable to smoke, flare, delay and other pyrotechnic composition. For black powder, a = 125 and b = 1300. Method for calculation of coefficient of friction is also enumerated in the paper[59].

For friction sensitivity, oblique impact tests are also developed. However, the outcome of these tests is like go-no-go states, where threshold values of sensitivity are ascertained. The process leading

---

[58] Jurie J.J.P.A. Le Roux, "The Dependence of Friction Sensitivity of Primary Explosives upon Rubbing Surface Roughness". *Propellant Explosives Pyrotechnics*, Vol. 15, No. 5, 1990, pp. 243–247.

[59] Duane M. Johnson, "Frictional Characteristics of Pyrotechnic Composition", 8th International Pyrotechnic Seminar, Colorado, USA, July 12–16, 1982, pp. 385–403.

to frictional heating and ignition of explosives needs proper care, attention and description. The control parameters like value of load, type of loading, orientation of loading, sliding speed, anvil roughness, roughness imparting material, substrate of the anvil etc are responsible for variation in friction sensitivity parameters and it can be used as a relative measure of sensitivity of compounds rather than a measure of their absolute hazards. In fact presence of grit between two surfaces enhances amount of heat generated by rubbing more heating occurs at the grit particles. The melting point of grit is responsible for maximum temperature realized and surprisingly grit hardness has no role in this. Grit can sensitize explosives, only when melting point of grits is higher than ignition temperature of the explosives. It is observed that no ignition is seen if metal file is rubbed against explosives. However placing some sand grit in between causes explosion. This grit is basically responsible for initiation of explosives by creation of local hot spots. Experimental measurement using a PBX (PBX 9501: 95% HMX, 2.5% Estane, 2.5% BDNPA/BDNPF) showed that grits tend to embed in the explosive and on frictional heating attains temperature to ignite explosive in around 0.1 seconds, when the onset of decomposition is observed[60].

Electrostatic spark sensitivity of explosives is another useful and important sensitivity parameter. It is defined as the degree of sensitivity of the explosive sample to high voltage discharge from a capacitor. In fact precise judgment of firing or initiation of explosive by an electrostatic discharge in a test chamber requires specialized equipments. Whether firing of explosive has taken place in the form of decomposition, burning or explosion is difficult. Sometimes, sound is taken as criteria, sometimes, flash or light is used, sometimes temperature is monitored and sometimes

---

60 Peter Dickson, Gary Parker, Laura Smilowitz, Jon Zucker, Blaine Asay, "Frictional Heating and Ignition of Energetic Materials", 36th International Annual Conference of ICT & 32nd International Pyrotechnics Seminar, Karlsruhe, Germany, June 29- July 1, 2005, p. 18–1.

61 Zhang Zhimin, Ling Fuxiu, "A New Method for Determining Electrostatic Sensitivity of Explosives and Propellants", Proceedings of the International Symposium on Pyrotechnics and Explosives, Beijing, China, Oct 12-15, 1987, pp. 859–865.

pressure is monitored to ascertain ignition. Advanced techniques like x-ray photography, electrostatic pulse methods are also used[61]. The energy from a capacitor discharge is given by $\frac{1}{2} CV^2$, where C = capacitance in Farad and V = voltage applied in volts. Configuration of electrodes, structure of circuit, time for discharge etc also affects numerical value of electric spark sensitivity values. This sensitivity parameter is related to square of detonation velocity, reciprocal temperatures, the activation energy and heat of fusion of explosives. A semilogarithmic relation between electric spark sensitivity and reciprocal of temperatures is also proposed. Without using density, heat of formation or detonation parameter, a simple correlation is developed for estimating electric spark sensitivity of explosives with molecular electronic properties. The correlation is given below.

$$E_{ES}\,(J) = (-1)^{n1} \times 10.16\, Q_{nitro} - 1.05\, n_1 n_2 E_{LUMO} - 0.20$$

Where, n_1 is number of aromatic rings, n_2 is number of substituted groups like alkyl or amino groups (other than nitro groups) attached to aromatic rings, Q_{nitro} (in e) is the minimum Mulliken charges of the nitro group given by sum of the charges of the 'N', 'O' atoms of nitro groups, E_{LUMO} (in eV) is the lowest unoccupied molecular orbit energy. If there is no substituted group other than nitro group to the aromatic ring, n_2 is taken unity.

For trinitro benzene, number of aromatic ring is 1 (n_1), number of substituted group is zero, and value of n_2) is taken as 1. The value of Q_{nitro} and E_{LUMO} are – 0.305 e and –3.850 eV respectively. So, E_{ES} = 3.0988 + 4.0425 – 0.20 = 6.9413 J, where as experimental value is 6.31 J. For TNT, values of n_1, n_2, Q_{nitro}, E_{LUMO} are 1, 1, – 0.326 e and –3.647 eV respectively. The value of electric spark sensitivity is 6.94 J (against experimental value of 6.85 J). The calculations are carried out for HNS, for which the values of n_1, n_2, Q_{nitro}, E_{LUMO} are 2, 1, –0.309 e and –3.904 eV respectively. The value of electric spark sensitivity is 4.86 J (against experimental value of 5.32 J). For relatively insensitive TATB also calculations are repeated. values of n_1, n_2, Q_{nitro}, E_{LUMO} are 1, 3, –0.490 e and –3.119 eV respectively. The value of electric spark sensitivity is 14.60 J (against experimental value of 17.75 J).

Initiation by Shock

Upon detonation, each explosive exhibits certain pressure in the detonation zone. If explosive mass is exposed to pressure higher than this pressure, it may be initiated. If pressure is less even then, explosives can be initiated, but this occurs with some appreciable delay. Shock pressure is basically a localized high pressure, which generates high temperature at the surface. This becomes real centers of initiation, which propagates or decays depending on the size of initial heated zone. If shock wave is initiated in explosive through barrier of different thickness, the shock first proceeds as non-reactive shock. This is followed by a detonating wave originating at the barrier surface or very close to it. This detonation starts in shock compressed explosive and propagates with a velocity considerably higher than ordinary steady state detonation. After a certain distance, the detonation catches up with the shock. The detonation decreases abruptly to a value higher than steady velocity and then slowly approaches to steady velocity. Just before first reaction, the explosive close to the barrier surface looses its transparency, which can be taken as first sign of onset of rapid chemical reaction. Detonation in first stage has high pressure and low temperature, so this gives less intense light. Any type of discontinuity or inhomogeneity in explosive by cavities, grit particles, grain boundaries or even grains of one explosive contained in another, result in early initiation (shorter delay time) at a given pressure. Inhomogeneous explosives can be initiated by pressures much less than that needed by homogeneous explosives. As pressure rises, depth of initiation of explosive from the barrier surface reduces.

When shock wave enters an explosive, it proceeds at accelerating velocity of lower magnitude. The acceleration depends on the shock pressure of incident wave. The velocity reaches detonation velocity in appreciable time. For steady detonation, conservation of momentum equation gives pressure as product of initial density, particle velocity and shock velocity in the explosive material. The space-time plot of this phenomena-using wedge of solid explosives gives a kink where shock is converted to detonation. Both shocked phase and detonation phase gives best fit curve as second order polynomial ($X = A + B \times t + C \times t^2$). In second order polynomial, constant term (A) has not much significance, as it only indicates 1-D spatial coordinates of initiation of shock and detonation fronts. The coefficient of linear term (B) indicates velocity and definitely, in detonation equation, this term has higher value than equation for shock. This is indirect way of statement that detonation velocity is higher than velocity of shock in the material. The quadratic term indicates acceleration. Compared to inhomogeneous explosives, relatively abrupt change to high-order detonation is observed in homogenous explosives.

There are evidences that after detonation, the non-conducting homogenous explosive becomes conducting. The shock initiation of homogenous explosives is more dependent on initial temperature and shock pressure than that for inhomogeneous explosives.

Shock initiated by a known donor charge is passed through brass attenuator of known thickness for assessment of shock pressure needed for initiation of explosive. The explosives of known thickness (also called receptor) are placed after attenuator and before plexiglass stack. The measurement of thickness of initiation is done after stack of plexiglass by suitable measurement equipment. The receptor charge thickness is varied to understand effect on detonation shock velocity. When there is no receptor charge, shock transmission from barrier has decaying nature with thickness. This gives reducing velocity of detonation with thickness. This is behaviour of a non-reacting surface or mass. When some receptor material is placed between barrier and plexiglass, the initial velocity of shock rises. This is due to contribution of detonating receptor in shock. The variation of shock velocity with thickness of plexiglass thickness shows a rising trend, indicating release of chemical energy from explosive behind the shock front. This increases pressure in the explosive with distance behind the shock front. As thickness of explosive charge is further increased, the rise in velocity reaches a maximum with thickness of plexiglass stack and then reduces. This indicates that energy released becomes closer to the shock front and on further increase in explosive thickness; the shock velocity reduces monotonically with distance.

Shock sensitivity of explosive is determined by placing barrier of different thickness between detonator and column of trial explosive. The perforation on a witness plate placed at the other end of the trial explosive column ensures detonation of explosives. The critical thickness at which detonation of trial explosive column is attained gives shock sensitivity of the trial explosive. Thickness of barrier is correlated with pressure part of shock pressure received from detonator and delivered at the other side to the trial explosive. The attenuation of different barrier materials has been tried for explosive shock. The barrier materials may be aluminium, PMMA (Poly-methyl methyl-acrylate) or cellulose acetate sheets. For Lucite rods[62] (considered similar to cellulose acetate sheets) attenuation of shock wave is studied and correlated with thickness of barriers.

Pressure, P (kbar) = $105 \times e^{-0.0358 \times thk}$, where thk = thickness of the attenuating barrier in mm.

For example, if critical barrier thickness of Lucite rod (cellulose acetate) is 20 mm, the shock sensitivity of the material is 51.31 kbar. This indicates that minimum 51.31 kbar pressure is needed for initiation of the explosive. It also indicates that for any thickness higher than 20 mm, the explosive will not be initiated, while at all barrier thickness less than 20 mm, it must initiate. For thickness of 10 mm, shock pressure delivered is 73.4 kbar. For TNT, composition B and Tetryl, shock sensitivity are calculated as 34.5 kbar, 19 kbar and 10 kbar respectively.

For initiation of secondary explosives charges, generally detonators are used which are highly sensitive for ease of initiation. The detonator quantity is kept bare minimum and relatively insensitive secondary explosive is initiated by the output shock energy of detonators. The explosive charges and detonators are generally prepared in the form of cylindrical charges. The height to diameter ratio of most of the detonators is commonly greater than 1 and sometimes upto 3. If explosive charge weight is constant,

---

[62] I. Jaffe, R. Beauregard, A. Amster, Determination of the Shock Pressure Required to Initiate Detonation of an Acceptor in the Shock Sensitivity Test, Proceeding of Third Symposium on Detonation, ONR Symposium Report, ACR-562, Vol. 2, Princeton University, USA, Sept 26-28, 1960, pp. 584–605.

an optimum ratio of height to diameter ratio of detonator is needed for effective initiation of main charge. Output of a detonator is expressed as $P^n \times t$, where P is pressure depending on type and density of the explosives and t is duration depending on the effectiveness of height of the detonator[63]. For a bare detonator, side rarefaction invades towards the centerline of the explosive column in detonator. The explosive column lying inside a cone with height equal to around diameter of the detonator is not affected by this side rarefaction waves. If explosive is contained in a rigid container, the effective length is 4/9 times length of the detonator charge because of the back rarefaction waves. This indirectly indicates that effective shape of the detonator is a right circular cone with diameter equal to diameter of the detonator. The greater the strength of the wall is, the longer the effective length of the charge is.

If shock is incident over an inert material, it will decay. However, if it is incident on an explosive, energy augmentation of shock wave occurs due to heat release by detonation of explosive. Thus decay of shock wave is arrested and shock wave propagates and strengthens in course of time. Prompt initiation occurs, if shock strength (in terms of pressure) and duration are sufficient. It is found that, if initiation ability of the detonator is insufficient, any one of diameter or height of the cylindrical detonator may be increased. However, it is practically observed that, after height is increased to certain values, it is more effective to increase diameter than to increase height. So there must be an optimum height, or an optimum height to diameter ratio for a given detonator. Based on a plate dent test, along with a side block, running distance and output dent is measured. The running distance decreases as height of detonator increases. After certain critical height, the running distance becomes constant. As height increases, dent depth increases and becomes constant after certain value of height is reached. Optimum height to diameter ratio is obtained between 0.8 and 1.2 to make full use of the detonator charge[64].

---

[63] Hayes D.B. "A Pnt Detonation Criterion for Thermal Explosion Theory", 6th International Symposium on Detonation, 1976.

[64] Cai Rui Jiao, Mao Jinsheng, "Optimum Ratios of Height to Diameter of Detonators", Proceedings of the International Symposium on Pyrotechnics and Explosives, Beijing, China, Oct 12–15, 1987, p. 11–116.

In fact secondary explosives have certain criterion for shock initiation. This is expressed by shock sensitivity, which is nothing but value of pressure, incident on explosives, which may lead to its initiation. The relation below gives the shock pressure for initiation of heterogeneous explosive.

P_{cr} (GPa) = $[(n-2)/(n-1)] \times \rho \times C^2 / (\gamma + 1)$, where n = pressure exponent of explosive burning, ρ = density of the explosive in g/cc, C = Shock velocity in km/s, γ = Gruneisen Coefficient of the explosives.

For Tetryl, n = 3, ρ = 1.65 g/cc, C = 2.36 km/s, γ = 1.5, hence shock sensitivity is 1.837 GPa. The practical value is found to be 1.85 GPa. Similarly, for PETN, n = 4, ρ = 1.60 g/cc, C = 1.32 km/s, γ = 1.5, hence shock sensitivity is 0.743 GPa. The practical value is found to be 0.95 GPa. For comp B-3, n = 4, ρ = 1.73 g/cc, C = 3.03 km/s, γ = 1, hence shock sensitivity is 5.2943 GPa. The practical value is found to be 5.63 GPa.

Shock sensitivity is generally measured by putting a barrier of calibrated material between a donor and test charge. The barrier delivers different shock pressures at the test-explosive. The barrier thickness can be calibrated to give shock sensitivity in terms of blast pressure directly. Different materials have different shock attenuating power and depending on material type, general correlations are developed[65]. The correlation depends on diameter of attenuating barrier also.

For aluminium alloy (Ly-12Al), shock velocity (U_s) is related to particle velocity (U_p) by relation $U_s = 5.48 + 1.30 \times U_p$. The dependence of shock velocity on thickness of barrier plate depends on barrier diameter. For higher diameter, shock velocity realized at a thickness is on higher side. This is because of rarefaction wave effects, which is predominant in smaller diameters. For smaller diameter (55 mm), Us (km/s) = $0.66 \sqrt{(99.53 - x)}$, where x is in mm. The distance can be measured in terms of time by the correlation x (mm) = $6.59 \times t - 0.11 \times t^2$, where t is time in seconds. A minor variation is observed, if aluminium barrier diameter is increased

[65] Chen Xirong, Wang Ke, Liu Derum, Wu Fengyuan, "Charaetristics of the Attenuation of Shock Waves in Barriers of Different Materials", Proceedings of the International Symposium on Pyrotechnics and Explosives, Beijing, China, Oct 12–15, 1987, pp. 875–881.

to 70 mm. U_s (km/s) = 0.68 $\sqrt{(100.2 - x)}$ and x (mm) = 5.80 × t – 0.12 × t^2. For Steel barrier in 70 mm diameter, U_s (km/s) = 0.72 $\sqrt{(49.62 - x)}$ and x (mm) = 5.06 × t – 0.13 × t^2. Under same load and at same distance, shock velocity is lower in steel. Different sound velocities in the barrier materials cause this difference. For steel $U_s = 3.57 + 1.92 \times U_p$. However for shock pressure, at same thickness, shock pressure in steel is highest but it decays at a faster pace. Shock pressure depends on material density, compressibility, shock velocity and particle velocity. Suppose PMMA (polymethylmethyl acrylate) is used as barrier material in 60 mm diameter size, $U_s = 2.87 + 1.88x\ U_p$. Other correlations are U_s (km/s) = 0.59 $\sqrt{(61.37 - x)}$ and x (mm) = 4.61 × t – 0.13 × t^2. The value of shock velocity in PMMA is lowest of the three barriers. Similar is trend for shock pressure, but decay of pressure is slower in PMMA.

If shock wave is to be accelerated to full detonation, delayed initiation occurs. The critical shock initiation pressure depends on production of hot spot in heterogeneous explosives. Frictional heating may create the hot spots. There are various other mechanisms of hot spot creation.

- Localized shear deformation[66,67]
- Heating caused by shock-induced rapid compression of gas spaces[68,69,70,71,72]

---

66 R.E. Winter, J.E. Field, "The Role of Localized Plastic Flow in the Impact Initiation of Explosives". *Proceeding of Royal Society, London*, A 343, 1975, pp. 399–413.

67 R.B. Frey, "The Initiation of Explosive Charges by Rapid Shear", Proceeding of 7th Symposium (International) on Detonation, *edited* by J.M. Short (Naval Surface weapons Center, White Oak, Silver Spring, MD, 1981, pp. 36–42.

68 R.B. Frey, "Cavity Collapse in Energetic Materials", Proceeding of 8th Symposium (International) on Detonation, *edited* by J.M. Short (Naval Surface weapons Center, White Oak, Silver Spring, MD, 1985, p. 68–80.

69 G.D. Coley, J.E. Field, "The Role of Cavities in the Initiation and Growth of Explosion in Liquids". *Proceeding of Royal Society of London*, A 335, 1973, pp. 67–86.

70 J.E. Field, "Hot Spot Ignition Mechanisms of Explosives". *Accounts Chemistry Research*, 25(11), 1992, pp. 489–496.

71 N.K. Bourne, J.E. Field, "Bubble Collapse and The initiation of Explosion". *Proceeding of Royal Society of London*, A 435, 1991, pp. 423–435.

72 F.P. Bowden, M.P. McOnie, "Formation of Cavities and Microjets in Liquids and Their role in Initiation and Growth of Explosion". *Proceeding of Royal Society of London*, A 298, 1967, pp. 38–50.

Calculation of shock initiation of explosives requires equation of state for explosives, product of explosion, unreacted explosives and their mixtures. Shock propagation through solid explosives occurs at a more or less constant or slightly decaying velocity. Shock heats the explosives. Behind the shock, reaction proceeds slowly at first and then accelerates as a result of selfheating. Initiation takes place after certain delay. The resulting detonation wave progresses through the heated explosive, which is also compressed by the already traversed shock. The velocity of this detonation wave surpasses shock velocity and after sometimes, it overtakes shock front. This gives over-driven detonation wave, which decays to steady detonation velocity subsequently. Shock initiation is investigated through mechanism of hot spot and discontinuities by most of the researchers.

Reactive flow model is the fundamental rate law for shock initiation of explosives[73]. In unreacted stage, explosives are solids, which on detonation get transformed into products, which are mostly gases. The model incorporates separate terms for ignition and propagation of reaction in explosives. The mechanism of hot spot creation by both the mechanisms described above (shear deformation and collapse of cavities) is included. The sequence of events is as follows.

- Creation of hot spots
- Coalescence of hot spots
- Subsonic deflagration
- Detonation of explosives

The ignition of explosives is attributed to thermal localization due to crystal inhomogeneity. Some crystals detonate even if there are no defects in them. PETN crystals have preferential direction for initiation. Along (110 and (001) planes, it can be detonated with 9 GPa stress, but for (100) and (101) planes, the threshold may go beyond 19 GPa. HMX is found to remain undetonated even when stressed by shock pressure exceeding explosive detonation

[73] John E. Reaugh, "Shock Initiation of Heterogeneous Explosives", Report No. UCRL-PROC-204120, International Conference New Models and Hydrocodes for Shock Wave Processes, College Park, MD, United States, May 12, 2004.

pressures. For such explosives, presence of defect is mandatory for initiation. Suppose defects are not present in HMX crystals, even a shock pressure beyond 400 kbar cannot detonate it.

In general, presence of defect leads to easier initiation of explosives by shock. However, in explosive formulations, rather than crystal defects, it is intra-granular porosity, which has higher magnitude. It is created by surface irregularities of the crystals, which are not filled completely by binders. If the defects and subsequently the hot spots are small, thermal conduction to surrounding cooler crystal removes heat faster than the continued decomposition replenishes it. So, there exists a minimum void size below which gas filled volume does not contribute to shock initiation.

The thickness of shock front is also responsible for initiation of explosives. Compared to void dimensions, shock front may be narrow or broad. For a narrow shock front compared to void dimension, void collapse is not as spherically symmetric quasi-static compression but as unsymmetrical collapse from a sharp shock. This unsymmetrical collapse of voids increases the chances of ignition for a given shock stress.

Reaction flow model advocates presence of isolated hot spots connected with each other by deflagration front to form a volume of hot products. The movement of deflagration front through unreacted solid depends on temperature and pressure. Although deflagration rates are subsonic, the quai-staic analysis can be conducted. Since the deflagration front size is very small (around 10 nm), stable time step is independent of whether explicit finite difference approach is using compressible hydrodynamic or heat transfer. The high pressure deflagration speed in HMX is calculated as , v (m/s) = $17 \times p^{0.52}$, where p is in GPa. The deflagration speed at low pressure is given by v (m/s) = $1.05 \times p^{1.726}$. The deflagration speed is dependent on temperature also.

Once initiated, the reaction propagates from hot spot. The hot spots are generally isolated and the cumulative behaviour is same as that by a single void. The final governing equation is as follows.

$dF/dt = 3 f_v^{1/3} (v/r_0) \times F^{2/3}$, where F is mass fraction of explosive consumed, f_v is volume fraction of defects and also the volume fraction of hotspots, v is propagation velocity, r_0 is radius of the hot spot.

Shock initiation of heterogeneous explosives are investigated by Forest fire model, reactive hydrodynamic code like SIN, 2DL, 2DE etc. Initiation by projectiles, sympathetic detonation, gap sensitivity test (transmission of explosive shock from donor to acceptor through attenuator), and optimization of booster explosive charge for given main charge etc is modeled successfully using these codes. Main mechanism for initiation by shock depends on charge failure diameter. If actual charge diameter is more than failure diameter, any detonation will propagate through the charge. Confinement reduces failure diameter. Sometimes shocks can desensitize explosive. This occurs when pressure is too low and too short to cause propagation of detonation through unshocked explosive charge from shocked explosive.

The mechanism of shock initiation of explosives may also be treated as transfer of detonation successfully from one explosive element to another. The propagation of detonation within explosives and between two explosives separated by certain distance (void/attenuator) is another dimension[74], which must be addressed after creation of hot spot and its conversion into full fledged steady detonation wave is ensured. The first part is initiation of an explosive. Each explosive has certain minimum or threshold amount of energy necessary to start the governing chemical reaction. This energy is responsible for hot-spot creation. This energy depends on chemistry of the materials, its density, confinement, pressure, temperature etc. This energy is quantified as integral of pressure-time profile and it depends on both magnitude and time duration of shock. Since booster, which passes detonation wave to secondary or main explosive, generally initiates explosives, the first element may be called donor, while another one may be called acceptor. The energy from donor and to acceptor

---

[74] S.S. Goldstein, D.L. Jackson, B.T. Nair, "Assessment of Current Methodologies to Establish Detonation Transfer Margins". AIAA 98-3467, Proceedings of 34th Joint Propulsion Conference, Cleveland, OH, July 1998.

for successful transfer of detonation is a function of its formulation, size, density and confinement of respective charges. In this connection gap testing or testing with some attenuator between both the charges is also significant.

TNT is one the most prominent explosives, which is investigated for its initiation characteristics. It is found that if density is maintained same, cast TNT is less shock sensitive than pressed TNT. This is because of fewer and larger holes in cast TNT than in pressed TNT resulting in less number of hot spots separated by longer distances. This is practical fact, which is confirmed by many researchers time and again. Similarly, it is also observed that pressed charges of bigger particle sizes (200-250 microns) are less shock-sensitive than pressed TNT-charges with finer particle sizes (20-25 microns). Such results are generally confirmed by either gap test or by wedge test and results in many cases are found contradictory also. Gap tests generally indicate coarser explosive more sensitive than fine because it senses a lower pressure region. Contrary to this wedge tests indicate reverse results[75]. The effect of particle size on shock sensitivity is investigated in RDX-PU based PBX also and formulations with fine RDX are found to be less sensitive at coarse-RDX formulations.

Hydrodynamic hot-spot model for initiation of explosives is further analyzed and it is obvious that explosives with coarser particle sizes have fewer holes or voids per unit volume resulting in fewer centers of hot spots. As explosives become finer, number of hot-spot locations increases but size of hot spots decreases. Smaller hot spot size results in cooling of high temperature hotspot center by side rarefaction before any appreciable decomposition of explosives take place. So shock sensitive reduces with reduction in particle-size. Another major concern is rate of reaction, which is numerically expressed as Arrhenius kinetics factors. If the reaction is fast enough to offsct the effects of side rarefaction, hot-spot decomposes before any cooling takes place. Explosives with faster Arrhenius kinetics exhibit higher shock sensitivity with decreasing particle sizes. The temperature before hot spot creation

---

[75] Donna Price, "Effect of Particle Size on the Sensitivity of Pure Porous HE". *Journal of Energetic Materials*, Vol. 6, No. 3–4, 1988, pp. 283–317.

also affects shock sensitivity and high temperature leads to earlier decomposition. An insensitive explosive may be made sensitive by elevation of temperature. TATB at 250°C has same sensitivity as HMX has at 25°C[76].

In a series of experiments with PBXs (Plastic Bonded Explosives), role of explosive molecules, its percentage, binder matrix is determined. The pressure needed to initiate the charge is ascertained by gap test. In this case tetryl (density = 1.56 g/cc, 57.5 mm dia × 50 mm height) is used as donor and attenuating media is PMMA (density = 1.186 g/cc and 70 mm diameter). The trial acceptor composition is made in 58 mm length and 58 mm diameter. The starting composition was HMX/HTPB (90/10). When 10 parts of HMX is replaced by Aluminium, the initiation pressure rises, making composition less sensitive. Further replacement of HMX, by aluminium powder leads to increase in insensitivity. In other words, if explosive ingredient is reduced, shock sensitivity reduces. Further trials showed that when HMX is replaced by RDX of higher percentage in a composition, impact sensitivity reduces. This indicates that compositions based on HMX are more sensitive than that made of RDX. So aluminized RDX based explosive composition gives a very insensitive explosive composition. Change in binder also affects sensitivity parameters. Macroscopic voids tend to increase sensitivity with increasing void diameter for sensitive compositions. However a relative insensitive explosive formulation become highly sensitive due to defects like debonding, grain cracks and local reactions caused by quasi-static and dynamic shock loads. This may enhance the risk of premature initiation in the hard target defeat penetrators with high velocities[77].

Visco-plastic heating is treated as effective mechanism in initiation of explosives. Viscosity and melting affects pore collapse.

---

[76] C.L. Mader, James D. Kershner, "Numerical Modeling of the Effect of Temperature and Particle Size of Shock Initiation Properties of HMX and TATB". *Journal of Energetic Materials*, Vol. 10, No. 2-3, 1992, pp. 69–95.

[77] Werner Arnold, "Significant Parameters Influencing the Shock Sensitivity Part I: Formulation and Binder, Part II: Voids and Defects", 36th International Annual Conference of ICT & 32nd International Pyrotechnics Seminar, Karlsruhe, Germany, June 29- July 1, 2005, pp. 188–1.

Melt layer at pores creates localized heating and rate of growth to detonation is determined by the rate of burning from the pores and grain boundaries. Maiden[78] has summarized sensitizing characteristics of explosives materials.

- Larger pores are facilitators for ignition
- Low conductivity also makes material sensitive
- High Arrhenius prefactor and low activation energy leads to sensitive compositions
- High activation energy of the melt is desirable for sensitive compositions
- Low viscosity and low melting points leads to faster initiation.

Through bulkhead initiation (TBI) is another method of initiation of explosives, where explosive charge is physically separated from detonator by an attenuating medium (metal), called bulkhead. This type of detonator is generally used for higher safety, as physically separation makes these charges less prone to accidental initiation. It has a donor charge and an acceptor charge separated by metallic bulkhead. When donor is initiated, shock wave passes through bulkhead and initiates charge located at other side of the bulkhead (acceptor), without rupturing bulkhead. In fact shock wave can be assumed to be propagating as compression wave and there is some form of energy dissipation while shock wave is propagating through metallic bulkhead. For a simplified one-dimensional analysis, for propagation of shock wave in stainless steel, following equations are valid.

$$d\sigma/dx = -24.82 \exp(-6.52 \times 10^3/\sigma T)$$

$$\sigma = \rho_0 U_p U_s$$

$$U_p = -3.00 + 0.66\, U_s$$

Where σ = stress of the shock wave in GPa, x = propagating distance in mm, U_p = Particle velocity in km/s, U_s = Shock wave

78 Dennis E. Maiden, "A Hot Spot Model for Calculating the Threshold from Shock Initiation of Pyrotechnics and Explosives". Proceedings of the Symposium on Pyrotechnics and Explosives, Beijing, China, Oct 12-15, 1987, pp. 594–610.

speed in km/s, ρ_0 = Initial density of stainless steel in g/cc, T = Temperature of environment in K.

The critical rupture stress in stainless steel plate with a thickness between 1.8 mm and 4 mm are experimentally determined as around 7.6 GPa (σ_{cr}). For design of such bulkhead, pressure at the interface of bulkhead and the acceptor charge is related to shock sensitivity of the explosive charge used and is given by the following expression.

$P_{int} = \frac{1}{2} \times (1 + \rho_m C_m / \rho_a C_a) \times P_{sse}$, where P_{int} = critical pressure at the interface, ρ_m = Density of metallic bulkhead, C_m = Shock velocity in metallic bulkhead, ρ_a = Density of acceptor charge, C_a = Shock velocity in acceptor charge, P_{sse} = Shock sensitivity of explosive.

The product of density and shock velocity in a medium is called shock impedance of the medium and in general, shock impedance of metallic bulkhead is on higher side as compared to that in porous acceptor medium. The shock wave attenuates as thickness of bulkhead increases in exponential fashion and at certain thickness; it attains pressure, which is insufficient to initiate the acceptor charge. Pressure (P_{sse}), corresponding to shock sensitivity of explosive gives higher limit of the bulkhead thickness and lower limit of interface pressure. Since bulkhead need not rupture, the lower limit of thickness of bulkhead is arrived at by the material strength property. As bulkhead thickness reduces, pressure at the interface rises and at certain lower thickness, it attains a value, equal to critical dynamic strength (σ_{cr}) of bulkhead material (for steel, it is 7.6 GPa). This gives lower limit of bulkhead thickness and higher value of pressure allowed.

The parameters, affecting the bulkhead design are derived from material characteristics, explosive quantity and properties and geometry of the bulkhead. Since pressing of granular charges generally makes acceptor charges, the initiation mechanism is invariably hot spot formation. For temperature less than 180°C, the shock sensitivity of acceptor charge may be taken as constant. Although dynamic rupture strength is also considered constant but it varies with temperature in not yet completely investigated manner. However, the decay of pressure in bulkhead with thickness depends on temperature and higher temperature gives higher slope for the pressure attenuation curve. In other words, if temperature is high, for a given thickness change; pressure change is significant, as compared to that at low temperatures. At low temperature, range of allowed bulkhead thickness is on higher side, but lower and upper limits of thickness both rises. There is less increase in lower limit of bulkhead thickness as compared to change in upper limit of bulkhead thickness due to temperature reduction. This changes the allowed thickness of bulkhead. If a bulkhead initiation mechanism has to work at a wide range of temperatures, the common allowed thickness range is to be selected[79].

Another approach for shock initiation of explosives may be practical determination of shocked explosives. In one of the report[80], PE4 (a white plastic explosive containing boiled and milled RDX grade B 88%, Grease 11% and pentaerythritol dioleate PEDO 1%) is considered for initiation by shock and it was found that – (*i*) high moisture content in composition causes failure, but larger plasticizer content do not lead to failure, (*ii*) Voids adjacent to detonator do not cause failure and (*iii*) void across complete diameter of sample can cause partial failure.

Particle size, particle morphology, surface defects and internal defects are responsible for varied shock sensitivity of explosives. The crystal defects as observed through NQR (nuclear quadrapole resonance) can be correlated with shock initiation pressures. Ultra-pure high quality RDX crystals are reported to give highly insensitive explosive formulations. Larger particle size (90-300 micron) of RDX has generally higher defect densities in the forms of cracks and pores than smaller particles (15-90 microns). Thus

---

[79] Dai Shizhi, Shen Ruiqi, "Estimation of the Designing Range of Bulkhead in a Through-Bulkhead Initiators", Proceedings of The International Symposium on Pyrotechnics and Explosives, Beijing China, Oct 12–15, 1997, pp. 22–27.

[80] M.C. Chik, L.A. Learmonth, "Determination of Shock Initiation and Detonation Characteristics of PE4 in Proof Test Geometries", Report No MRL-A-979 Material Research Laboratories, Melbourne, Australia, Nov 1985.

formulations using higher particle sizes have marginally lower shock initiation pressure[81].

Initiation of explosives by shaped charge jet also forms part of this category. Shaped charge has lined cavity, which on initiation creates small diameter high velocity projectile. The strength of jet (velocity) is reduced by passing it through steel attenuator, before any interaction with explosive. Critical jet value of initiation of explosive by jet is given by product of jet diameter (d) and square of jet velocity (V^2). Various values of V^2d are possible for combination of liner material and explosives[82].

Velocity is recorded in mm/ms and diameter is expressed in mm, generally. The value of V^2d is expressed in $mm^3/\mu s^2$. For RDX/TNT 60/40, at density of 1.70 g/cc, the critical parameter for the copper jet is 5.8 $mm^3/\mu s^2$. Any value of critical parameter higher than this value will lead to initiation of the charge. For PBX- 9502 (TATB/ Kel-F 95/5 at 1.894 g/cc density), the minimum value of critical parameter is 127 $mm^3/\mu s^2$, for copper jet (jet diameter = 3.80 mm) initiation. The C-J Pressure for PBX-9502 is 285 kbar and its failure diameter in unconfined case is 9 mm. However, copper jet of less than 4 mm is sufficient to initiate PBX-9502 due to high pressure encountered. For copper jet at 5.0 mm/ms shocking inert PBX-9502, the pressure match is reported as 680 kbar which improves to 920 kbar for copper velocity of 6.0 mm/μs.The shock pressure sent into the explosive material depends on jet material. Aluminium has lower density than copper. So, aluminium jet must possess greater diameter and/or velocity than copper jet for equal impulse on PBX-9502. So, V^2d for aluminium is higher than that for copper jet and practically, it is found to be 325 mm^3/ms^2 as compared to 127 $mm^3/\mu s^2$ for copper jet. Since water jet has further reduced density, the value of minimum critical parameter for PBX-9502 is 800 $mm^3/\mu s^2$ for water jet initiating the charge. For 75/25 cyclotol (RDX/TNT), the value is 37 $mm^3/\mu s^2$, for copper jet initiation.

---

81 S.M. Caulder, P.J. Miller, K.D. Gibson, J.M. Kelley, "Effect of Particle Size, Particle Morphology and Crystal Quality on the Critical Shock Initiation Pressures of Cast RDX/HTPB Compositions", 37th International Annual Conference of ICT, Karlsruhe, Germany, June 27–30, 2006, p. 9–1.

82 C.L. Mader, G.H. Pimbley, "Jet Initiation of Explosives", Report number - LA-8647, Los Alamos Scientific Laboratory, New Mexico, Feb 1981.

Initiation of explosive by shaped charge jet is different from that by other mechanisms. In general, for initiation, shock of sufficient strength (pressure or velocity) or duration builds up detonation. Propagation of detonation is assured in this case by large geometry (larger than failure diameter). However, in near critical jet initiation, prompt detonation of explosive propagates only if shock wave produced by jet is of sufficient magnitude and duration. So, any jet may produce significant amount of energy from an explosive even below the critical value of V^2d required for propagation of detonation, but stable initiation of explosive may not occur.

Initiation by Other Means

Shock initiation of explosive desires shock pressure of sufficient strength to act for sufficient duration on the explosive. However, in case of impact of thin (3 μm) flyers on explosives, very small duration of impact of the order of 1 nanosecond is sufficient to initiate the explosive. Hot spot mechanism discussed above must be able to release energy in this short duration. In this case shock thickness is similar to hot-spot sizes and less than grain size also. When explosive is initiated by projectile impact, mechanism can be though as initiation by a shock of some equivalent pressure. If projectile and explosive charge, both are cylindrical with plane end surface and the projectile

Impact pressure, $p = \rho_e w_e v/[1+\rho_e w_e/\rho_p w_p]$, where v = projectile velocity before impact in m/s, ρ = density in g/cc, w = shock wave velocity, subscript e = explosive and subscript p = projectile.

The flyer plate initiation of explosives depends on explosive material, grain size of explosive and so on. There are many interesting results on the parametric variations. For HNS crystals, it is observed that low surface area or bigger particle size makes it more sensitive to shock initiation[83]. Reverse trend is also

[83] R.E. Setchell, "Experimental Studies of Chemical Reactivity During Shock Initiation of Hexanitrostilbene", Proceeding of Eighth Symposium (International) on Detonation, *edited* by J.M. Short (Naval Surface Weapons Center), White Oak, Silver Spring, MD, 1985, p. 15–25.

observed[84]. 1,3,5-tri-nitro benzene is found more sensitive in finer particle sizes for flyer plates. Whatever may be experimental observation and their probable reason, it is clear that grain size is one of the governing parameter for explosive initiation by flyer plates impact. For PETN, it is observed that if grains are fine, it is initiated at lower flyer impact velocities. This means fine grain PETN is more sensitive[85]. It is clear that fine particle sizes give a smoother surface finish than coarser particle sizes. Before decaying, shock waves will traverse many layers and voids in finegrained PETN than in coarse-grain PETN. For coarse-grain PETN shock will impact only at the high points. In both the cases airvoids will collapse rapidly producing hot-spot either by gas heating or jet impact. The size of hot spot is of the order of microns. For coarser grain void or inter-surface distance are of higher order, which leads to decay of hot spot effectiveness. The transfer of heat to adjacent surface is restricted and rise in bulk-temperature for initiation of exothermic reaction becomes sluggish in coarsegrain PETN. Contrary to this fine-grain PETN undergoes more reaction due to higher specific surface area during the shock transition period. In addition to this fine grain materials are stronger and are less likely to fail by brittle fracture. They undergo plastic failure by localized shear banding. However initiation and propagation of shear bands should be fast enough to cause initiation of bulk explosive.

Plate Flyer velocity can be treated as an important parameter for indicating explosive energy. However, it is not a constant for given explosive. It depends on properties of explosive, experimental conditions, amount of explosives, geometry of explosive, flying distance, presence or absence of confinement etc. There are many tests like flat plate test, sphere test, cylinder test to determine flyer plate velocity. In a typical test set-up, where explosive charge in 100 mm diameter and 80 mm thickness was initiated by a plane wave lens. A 2 mm thick steel plate was driven

---

[84] D.B. Hayes, "Shock-Induced Hot-Spot Formation and Subsequent Decomposition in Granular, Porous HNS Explosive". *Progress in Astronautics and Aeronautics*, 87, 1983, pp. 445–467.

[85] S. Watson, M.J. Gifford, J.E. Field, "The Initiation of Fine Grain Pentaerythritol Tetranitrate by Laser Driven Flyer Plates". *Journal of Applied Physics*, 88(1), July 2000, pp. 65–69.

by detonation products. As flyer traveled 45 mm distance, the instantaneous velocity was recorded by streak camera. An empirical formula is developed for prediction of flyer plate velocity[86].

Flyer plate velocity, V_{fp} (km/s) = $(0.675 - 0.054 \times K) \times D + 0.56 \times \rho - 1.039$, where K = polytropic index in the equation of state of detonation products, measured by free surface velocity of metal plates, D = detonation velocity in km/s, ρ = density of explosive in g/cc.

Laser initiation is also used for explosive, but with certain limitations. Although laser initiation offers high safety, reliability and maintainability, the progress on application of laser initiation of explosive is rather slow. Laser initiation of pyrotechnic mixtures was reported around 50 years ago[87]. Later on initiation of secondary explosives like PETN, RDX, Tetryl etc were also tried for initiation by laser. Experimental data reveal that PETN has lower threshold initiation energy, shorter induction period and stable performance among high explosive. This makes it an attractive candidate for initiation by laser sources. The mechanism of initiation by laser may be treated through thermal or shock. Thermal means of propagation may lead to deflagration to detonation transition and a relatively longer induction period. Adding Zr-powder and forming a low-density mixture reduces induction period. PETN cannot be initiated with laser beam diameter less than 0.8 mm. As diameter of laser beam is increased, laser energy and density needed to initiate explosive reduces. For fine crystalline PETN, critical beam diameter coincides with critical diameter of explosive (0.70–0.86 mm). The critical laser energy density for initiation of PETN is 3.3 J/cm^2, which corresponds to a laser energy of 73 mJ[88].

---

86 Hua Pinghuan, "Prediction of Plate Flyer Velocity", Proceedings of the International Symposium on Pyrotechnics and Explosives, Beijing, China, Oct 12–15, 1987, pp. 911–915.

87 Lewis D.J., "Laser Energized Explosive Device System", Proceedings of 5th Symposium on Electro-Explosive Devices, 1969, p. 1–8.

88 Jia Baoren, Chen Danming, Wang Zuoni, Li Yulian, "Experimental Studies on PETN Detonators Initiated by Laser Radiation", Proceedings of the International Symposium on Pyrotechnics and Explosives, Beijing, China, Oct 12–15, 1987, pp. 49–56.

7

Prediction of Detonation Parameters

Detonation parameters

Detonation of explosives is always expressed in terms of three important parameters. Velocity of detonation, ratio of specific heats and detonation pressure are the parameters, which are considered as benchmark for explosive performance. Velocity of detonation is velocity with which shock wave augmented by chemical reaction progresses through the explosive on suitable initiation. It is constant for an explosive. But density of explosives affects it. Higher density gives higher velocity of detonation. There exists a minimum diameter above which this velocity of detonation (VOD) is constant. Below this diameter, VOD increases with increase in explosive charge diameter. Ratio of specific heat is a theoretical terms for explosives and contrary to normal range from 1 for solids and liquids to 1.67 for mono-atomic gases, for detonation products, it has much higher values. For one of lead based explosive, the value is reported to be as high as 6. This is due to high pressure and temperature conditions at the explosion sites. The detonation pressure is correlated to detonation velocity, density and ratio of specific heats. It is highest pressure recorded at the point of detonation.

Prediction of detonation velocity for explosive molecules is one of the major challenges. Theoretically arriving at practically measured detonation parameters suffers from many uncertainties. On theoretical front, the mechanism for breaking of explosive molecules is not known. The chemical reaction kinetics, reaction rate parameters, mechanism of dissociation, temperature and pressure dependent performance, unavailable equation of states are some of the parameters, which prevents accurate estimation of detonation parameters theoretically. On the other hand, practical determination of detonation parameters also suffers from inaccuracies in measurement system, response time of sensors, transient phenomena, spatial and temporal variation of parameters, high pressure and temperature conditions etc. Overall a compromise is made to understand the behaviour and at most a relative higher or lower performance of materials can be ascertained. However, exact determination or calculation of detonation parameters is not possible despite high-speed computational facilities, advanced instrumentations and progress in theoretical understanding. For calculation of velocity of detonation of explosives, many computer codes are developed and used. RUBY[89], TIGER[90], CHEETAH[91], BKW[92] are computer codes, which are developed for such applications. But they need many input parameters for the given output and are a black box for the users. Preparation of input file, requirement of a computer, cost of the codes and other requirements make these codes not very user-friendly for quick calculations. However, many empirical attempts are made to theoretically predict detonation parameters from molecular formula, density and heat of formation of explosive molecules. This section covers some of the empirical relations used for performance prediction of explosives.

Another important detonation parameter is ratio of specific heats (γ) for detonation products for the explosive molecules and composition. Generally for explosives, value of γ varies between 2.5 and 3.3 and average value is taken as 3. However there are many formulations available in open literature, which gives γ as a

[89] H.B. Levine, R.E. Sharples, Operator's Manual for RUBY, Lawrence Livermore Laboratory report UCRL-6815, Livermore CA, 1962.

[90] M. Cowperthwaite, M.W.H. Zwister, TIGER Computer Program Documentation, Stanford Research Institute, SRI Publication number 2106, 1973.

[91] L.E. Fried, W.M. Howard, P.C. Souers, CHEETAH 2.0 user's manual, Lawrence Livermore National Laboratory report , Livermore CA, 1998.

[92] C.L. Mader, "Numerical Modeling of Explosives and Propellants", Second Edition, CRC Press, 1998.

function of density of explosive molecules or explosive compositions. Some of the relations are given below.

$$\gamma = 1.90 + 0.6 \times \rho$$
$$\gamma = \rho / (0.140 + 0.260 \times \rho)$$
$$\gamma = \rho / (0.136 + 0.238 \times \rho)$$
$$\gamma = (0.655/\rho) + 0.702 + 1.107 \times \rho$$

Another detonation parameter is detonation pressure, which is calculated from density, ratio of specific heats of detonation products and velocity of detonation of explosive. A common relation is $P\ (\text{kbar}) = 10^{-5} \times \rho\ (\text{g/cc}) \times D^2\ (\text{m/s})/(1 + \gamma)$.

Kamlet and Jacob Method

Jacob and Kamlet[93] gave the oldest relation used for performance prediction of explosives. This is valid for explosives containing carbon, hydrogen, nitrogen and oxygen molecules. The method relies on determination of three parameters for any trial molecule given by chemical formula $C_aH_bN_cO_d$. The first parameter is chemical energy released or Q, which is already calculated in earlier section for individual molecules. In fact, if equivalent weights are available, the same can be calculated for a mixture of explosives, also. Second parameter is moles of gaseous detonation products per gram of the explosive molecules. 'N' denotes this parameter. Third parameter is called average molecular weight of gaseous product of explosion and it is denoted by 'M'. It is calculated by dividing total molecular weight of gaseous product of explosion by number of moles of gaseous products formed.

Generally, explosives are oxygen deficient molecules or mixtures. So, they do not have sufficient oxygen to oxidize all the hydrogen and carbon available. As per conversion schemes discussed earlier in prediction of single molecules section, solid carbon is not taken as gaseous product of combustion. A general chemical equation can be written as follows:

---

93 Mortimer J. Kamlet, S.J. Jacobs, "Chemistry of Detonations I. A Simple Method for Calculating Detonation Properties of C-H-N-O Explosives". *The Journal of Chemical Physics*, Vol. 48, No. 1, Jan 1968, pp. 23–35.

$$C_aH_bN_cO_d \rightarrow (c/2)\ N_2 + (b/2)\ H_2O + (d/2 - b/4)\ CO_2 + (a - d/2 + b/4)C$$

From the equation, it is clear that number of moles of gaseous molecules is given by (c/2) + (b/2) + (d/2–b/4) = ¼ (2c + 2d + b). With this, the value of 'N' is calculated as given below:

N = ¼ (2c + 2d + b) / molecular weight of the explosive molecule = ¼ (2c + 2d + b)/(12a + b + 14c + 16d)

The value of 'N' for RDX ($C_3H_6N_6O_6$)
= ¼(12 + 12 + 6)/222.174 = 0.0337573

For the scheme of chemical reaction given above, value of 'M' can also be calculated. The molecular weights of nitrogen gas, water and carbon di-oxide are 28, 18 and 44 respectively. Molecular weight of product of explosion and their number of moles give value of M.

$$M = [28 \times c/2 + 18 \times b/2 + 44 \times (d/2 - b/4)]/[¼(2c+2d+b)] = (56c + 36b + 88d - 44b)/(2c + 2d + b)$$
$$= (56c + 88d - 8b)/(2c + 2d + b)$$

The value of 'M' for RDX ($C_3H_6N_6O_6$)
= (56 × 6 + 88 × 6 – 8 × 6)/(12 + 12 + 6) = 27.2

Kamlet and Jacob proposed an intermediate term $\phi(=N\sqrt{M}Q)$ to take care of the three parameters in totality. The value of detonation velocity (D) is estimated by the formula given below:

$$D \text{ (in km/s)} = 1.01(1 + 1.3\rho)\sqrt{\phi}$$

For RDX
Q = 1481 cal/g
N = 0.0337573
M = 27.2
ϕ = 6.7754
$D = 1.01\ (1 + 1.3 \times 1.805)\ \sqrt{6.7754} = 8.8013$ km/s

Kamlet and Jacob have given another empirical relation for calculation of detonation pressure.

P (kilobar) = 15.58 × ρ^2 × ϕ

For RDX,

P = 15.58 × 1.805^2 × 6.7754 kbar = 344.302 kbar = 34.4302 GPa.

Value of ratio of specific heat from the relations by Kamlet for velocity of detonation (D) and detonation pressure (P) given can be calculated as $\gamma = (10^{-5} \times \rho \times D^2/P) - 1 = [(10^{-5} \times 1.805 \times 8801.3^2) / 344.302]/-1 = 3.061$.

All the detonation parameters calculated as per Kamlet prediction scheme for the explosive molecules given in Table 1, is tabulated in Table 7. However, the calculations are based on arbitrary assignment of product of explosion. There are other schemes for arriving at the product of explosion also and some of the schemes for RDX are given below.

Scheme 1 : $C_3H_6N_6O_6 \rightarrow 3N_2 + 3H_2O + 1.5\ CO_2 + 1.5\ C$
Scheme 2 : $C_3H_6N_6O_6 \rightarrow 3N_2 + 3CO + 3H_2O$
Scheme 3 : $C_3H_6N_6O_6 \rightarrow 3N_2 + 3CO_2 + 3H_2$

Table 7: Detonation parameters of explosives by Kamlet

Explosives	*N*	*M*	*Phi*	*D (m/s)*	*P (kbar)*	*Gamma*
RDX	0.03376	27.2	6.77543	8801.35	344.302	3.06328
TNT	0.02531	28.522	4.86288	7016.27	207.268	2.9284
HMX	0.03376	27.2	6.76242	9130.92	382.349	3.15398
PETN	0.03163	30.4	6.78306	8717.4	334.837	3.0398
Tetryl	0.02699	30.452	5.61002	7772.37	261.592	2.99511
HNS	0.02332	32	4.87489	7274.24	229.949	3.004
NQ	0.03842	23	5.47718	7925.62	279.564	3.0669
Explosive D	0.02844	28.286	4.98365	7287.52	228.905	2.98359
TATB	0.02905	27.2	4.96621	7921.41	290.603	3.18465
DATB	0.02776	28.444	5.07098	7705.9	266.611	3.09146
NTO	0.03075	29.5	5.42824	8257.23	315.022	3.17719

Calculation as per scheme 1 is already elaborated in detail above and velocity of detonation for RDX is estimated as 8.8013 km/s. However, by other schemes different values are possible.

Scheme 2: $C_3H_6N_6O_6 \rightarrow 3N_2 + 3CO + 3H_2O$

$N = (3 + 3 + 3)/222.174 = 0.0405087$

$M = (3 \times 28 + 3 \times 28 + 3 \times 18)/9 = 24.66666$

$Q = (14.71 + 3 \times 26.46 + 3 \times 57.79)/222.174$ kcal/g = 1203.83 cal/g

$\phi = N\sqrt{MQ} = 6.9804934$

$D = 1.01\,(1 + 1.3 \times 1.805)\,\sqrt{6.9804934} = 8.930$ km/s

Scheme 3: $C_3H_6N_6O_6 \rightarrow 3N_2 + 3CO_2 + 3H_2$

$N = (3 + 3 + 3)/222.174 = 0.0405087$

$M = (3 \times 28 + 3 \times 44 + 3 \times 2)/9 = 24.6666$

$Q = (14.71 + 3 \times 93.98)/222.174$ kcal/g = 1335.215 cal/g

$\phi = N\sqrt{MQ} = 7.351553$

$D = 1.01(1 + 1.3 \times 1.805)\,\sqrt{7.351553} = 9.1643$ km/s

The various parameters are compiled in tabular form (Table 8) to understand the cause of such variations in prediction.

Table 8: Detonation velocity calculation by Kamlet's method

Scheme	*N*	*M*	*Q*	*ϕ*	*D*
			cal/g		*m/s*
1	0.03376	27.2	1481	6.7754	8801
2	0.0405087	24.667	1204	6.98050	8930
3	0.0405087	24.667	1335	7.35155	9164

Clearly, the value of detonation velocity changes with scheme selected for estimating product of explosion. In scheme 1, formation of solid carbon (C) is envisaged, which reduces number of moles of gaseous products. This solid carbon does not contribute by any means in the chemical heat release also. But since complete combustion releases more heat, both hydrogen and remaining carbon gets completely oxidized to water and carbon dioxide, resulting in higher release of energy. However, rise in energy is not sufficient to compensate the loss in number of moles of gaseous molecules, resulting in small value of Kamlet parameter 'ϕ'. Low value of this parameter at a given density of explosive will give lower velocity of detonation (VOD). Between schemes 2 and 3, it is clear that number of moles of gaseous products produced and molecular weight of product of explosion are same. In scheme 3, all carbon is completely oxidized, while in scheme 1, all hydrogen is completely oxidized. These are two extreme cases. Since heat release per atom of oxygen is more, if hydrogen is oxidized, chemical heat release in scheme 1 is found more. Chemical heat release is less in scheme 3 than in scheme 1. However, Kamlet parameter 'ϕ' has much higher value as per scheme 3. High value of this parameter at a given density of explosive will give higher velocity of detonation (VOD). In case of scheme 2, hydrogen is completely oxidized, but carbon is partially oxidized. This will have the lowest heat release than all other schemes. However, higher number of moles of gaseous products than scheme 1, leads to higher 'f' and D by scheme 2 than by scheme 1. If experimental measurement of VOD for RDX at a given density is compared with calculated results, scheme 1, is found to be most suitable way of calculating VOD. Kamlet's method is invariably used for prediction of detonation parameters of explosives reliably. Another major advantage of this method is extension of its usages to the domain of explosive formulations.

The scheme can be extended to explosive formulations also. To show the relevance of this calculation scheme by Kamlet for explosive formulation, first explosive formulation containing 75% by weight RDX and 25% by weight TNT is considered. The equivalent formula as calculated earlier is $C_{3.9824}\,H_{5.7548}\,N_{5.2615}\,O_{6.000}$ and molecular weight of the formulation is 223.23 g/mol as calculated. The heat of formation on weight average basis is calculated to be 3.317 kcal/mol and density of the formulation is 1.75 g/cc. The value of chemical heat release by detonation of the gas is calculated as 1416.2 cal/g as per scheme 1 discussed earlier. This is equivalent to 2478.34 cal/cc heat release. In scheme 1, after formation of water, remaining oxygen is utilized in formation of carbon-di-oxide directly. As per second scheme, carbon is first oxidized to carbon mono-oxide and the remaining oxygen oxidizes carbon mono-oxide to carbon di-oxide. As per second scheme chemical energy release is calculated to be 1129.18 cal/g, which is equivalent to 1976.06 cal/cc. For detonation parameter by method suggested by Kamlet, N and M are calculated to be 0.031099 and 27.46385 respectively for the explosive composition. As per scheme 1 of chemical energy release, value of ϕ = 6.133185.

$D = 1.01\ (1 + 1.3 \times 1.75)\ \sqrt{6.133185} = 8.214$ km/s

$P\ (\text{kilobar}) = 15.58 \times \rho^2 \times \phi = 15.58 \times 1.75^2 \times 6.133185$

$= 297.77$ kbar

$\gamma = [(10^{-5} \times 1.75 \times 8191.7^2)/292.63] - 1 = 3.026.$

Some more treatments are given to calculation of detonation parameters or explosive mixtures by a paper by Zhou Xing Xi *et al.*[94]. Three methods are proposed in the paper for the calculation of Kamlet parameter, ϕ for the explosive mixtures. As per first method, equivalent composition is calculated, from which value of 'N'. 'M' and 'Q' are calculated and then ϕ for the equivalent composition is calculated. For a 50:50 RDX/TNT composition, equivalent composition is $C_{4.98}H_{5.505}N_{4.517}O_6$, with a molecular weight of 224.638 g/mol and theoretical maximum density of 1.7312 g/cc. The equivalent heat of formation is 0.815318 kcal/mol. The calculations in the paper for this composition are reported at a density of 1.627 g/cc and velocity of detonation experimentally achieved is 7.66 km/s. The probable chemical reaction during detonation is given below.

$$C_{4.98}H_{5.506}N_{4.517}O_6 \rightarrow 2.2585\ N_2 + 2.753\ H_2O + 1.6235\ CO_2 + 3.3565C$$

$N = (2.2585 + 2.753 + 1.6235)/224.638 = 0.029535$

$M = (2.2585 \times 28 + 2.753 \times 18 + 1.6235 \times 44)/6.635$

$= 27.766$

$Q = (0.815318 + 1.6235 \times 93.98 + 2.753 \times 57.79)/224.638$ kcal/g $= 1391.059$ cal/g

$\phi = N\sqrt{MQ} = 5.804573$

$D = 1.01\ (1 + 1.3 \times 1.627)\ \sqrt{5.804573} = 7.580$ km/s

The above-mentioned scheme is more calculation intensive and other methods are also proposed in the paper. Another scheme proposed for calculation of Kamlet parameters is relatively easier,

[94] Zhou Xing Xi, Yu Yong Zhong, "Analysis on Computation of Kamlet Parameter f for CHNO Explosive Mixtures". *Journal of Energetic Materials*, Vol. 9, No. 4, 1991, pp. 283–296.

where weight-average values of 'N', 'M' and 'Q' are considered for calculation of final 'ϕ' of the composition.

$$N_{comp} = \Sigma N_i w_i = 0.50 \times 0.033757 + 0.5 \times 0.025313 = 0.029535$$
$$M_{comp} = \Sigma M_i w_i = 0.50 \times 27.2 + 0.5 \times 28.522 = 27.866$$
$$Q_{comp} = \Sigma Q_i w_i = 0.50 \times 1481 + 0.5 \times 1294 = 1387.5 \text{ cal/g}$$
$$\phi = N\sqrt{MQ} = 5.8144$$
$$D = 1.01\,(1 + 1.3 \times 1.627)\,\sqrt{5.8144} = 7.586 \text{ km/s}$$

Another simpler reported method is considering weight-average value of Kamlet parameter 'ϕ' for calculation of the same for the given composition.

$$\phi_{comp} = \Sigma \phi_i w_i = 0.50 \times 6.7887 + 0.5 \times 4.8653 = 5.82698$$
$$D = 1.01\,(1 + 1.3 \times 1.627)\,\sqrt{5.82698} = 7.595 \text{ km/s}$$

All the methods give velocity of detonation very near to the experimentally observed values. However, the values are closer for the third scheme, where fortunately calculation is easier also. The paper compiles data for several explosive compositions and concludes that third scheme in fact gives best approximation for the explosive mixtures (based on regression coefficients).

Table 9: VOD of RDX/TNT compositions

RDX	*TNT*	*Density (g/cc)*	*VOD (km/s)*			
			EXPTL	*Method I*	*Method II*	*Method III*
0	100	1.654	N.A.	7.018	7.018	7.018
50	50	1.627	7.66	7.58	7.587	7.595
60	40	1.72	7.89	8.004	8.011	8.019
64	36	1.713	8.018	8.033	8.04	8.047
65	35	1.715	8.06	8.053	8.059	8.066
75	25	1.648	7.952	7.961	7.965	7.971
77	23	1.754	8.25	8.336	8.34	8.346
78	22	1.755	8.306	8.352	8.357	8.362
100	0	1.816	N.A.	8.844	8.844	8.844

The calculations by all three schemes are repeated for various combinations of compositions of RDX/TNT and the results of calculations are reproduced in Table 9. For completeness, calculations are repeated for pure explosives RDX and TNT also by all the three methods. For pure explosives, all three schemes give same value of VOD. However, for explosive compositions, the values vary depending on method adopted.

Although, there are many assumptions and routes for estimating final product of explosion and the method of calculation of Kamlet parameter 'ϕ' after finalization of final products also has several routes, the final output is always found very near to the actually observed values of VOD for explosive compositions. There are several parameters, which are correlated to this Kamlet parameter and the relations are given below in summary form.

Gurney velocity, $\sqrt{2E}$ (km/s) $= 0.53 \times \sqrt{(1.44 \times \rho \times \phi)} + 0.60$

Gurney velocity, $\sqrt{2E}$ (km/s) $= 0.887 \times \sqrt{(\phi\sqrt{\rho})}$

Cylinder velocity, V(km/s) $= 0.368 \times \rho^{0.84} \times \phi^{0.54}$

Cylinder energy, E (MJ/kg) $= 0.0659 \times \rho^{1.8} \times \phi$ for 6 mm

$= 0.0927 \times \rho^{1.6} \times \phi$ for 19 mm

Relative detonation impulse, $I = 15.06 \times \rho^{1.182} \times \phi^{0.725}$

Adiabatic exponent, $\gamma = 184\,(N^2M/\phi) \times \rho + 1.168$

The Kamlet parameter is used for prediction of cylinder wall velocity and cylinder energies at 2 ($E^{6\ mm}$) and 7 ($E^{19\ mm}$) volume expansion also. Following correlations are developed for such calculations[95].

$$V_{cyl\ wall}\ (km/s) = 0.368 \times \phi^{0.54} \times \rho^{0.84} \times (R - R_0)^{0.212-0.65\rho}$$

$$E^{6\ mm}\ (MJ/kg) = 0.0659 \times \phi \times \rho^{1.8}$$

$$E^{19\ mm}(MJ/kg) = 0.0927 \times \phi \times \rho^{1.6}$$

[95] Ruth M. Doherty, James M. Short, Mortimer J. Kamlet, "The Chemistry of Detonations. X. Prediction of Cylinder Test Energies of Usual and Non-usual Explosives", Proceedings of the Symposium on Pyrotechnics and Explosives". Beijing, China, Oct 12-15, 1987, pp. 489–496.

$$(V_{20})^2 = 0.169 \times \phi^{1.095} \times \rho^{1.48}$$

$$(V_{20}/V_6)^2 = 1.919 - 0.331 \times \rho^{1.6} - 0.39\ [MF_{H2O}]^2$$

Several researchers tried to validate and modify relations proposed by Kamlet. In a paper, Keshavarz[96] proposed modification to the method proposed by Kamlet. The developed formulation is claimed to be valid for explosives with loading densities less than 1 g/cc. At lower loading densities, Kamlet's formulations predict higher than experimentally observed values of velocity of detonation. To overcome this lacuna, while maintaining versatility of Kamlet's formlation for higher loading densities, a modification to Kamlet's original method is suggested and the correlation is given below.

$$D\ (km/s) = (0.779 + 1.442 \times \rho)\ \sqrt{\phi}$$

The Kamlet's equation is evaluated against experimental data in open literature. One of the reported compositions is a PBX made polyurethane (Estane) foam and PETN in the ratio 40/60[97]. The molecular formula for Estane is given as $C_{5.14}H_{7.50}N_{0.19}O_{1.76}$. It has density of 0.8744 g/cc and heat of formation of –95 kcal/mol. PETN is $C_5H_8N_4O_{12}$, with density of 1.76 g/cc and heat of formation of –128.74 kcal/mol. The velocity of detonation experimentally evaluated as 3600 m/s, 3970 m/s and 5541 m/s at densities of 0.5 g/cc, 0.6 g/cc and 1.0 g/cc, respectively. To apply Kamlet's method, equivalent formula is calculated first as $C_{5.09}H_{7.66}N_{1.42}O_{5.06}$. The molecular weight is 169.71 and heat of formation is –62.376 kcal/mol. The proposed chemical reaction by Kamlet's method is given below.

$$C_{5.09}H_{7.66}N_{1.42}O_{5.06.} \rightarrow 0.71\ N_2 + 3.83\ H_2O + 0.615\ CO_2 + 4.475C(s)$$

N = 0.030357829 mol/g of explosive

---

[96] M.H. Keshavarz, "Prediction of the Principal Performance Parameters of Explosives Over a Wide Range of Initial Densities". *Asian Journal of Chemistry*, Vol. 17, No. 4, 2005, pp. 2085–2092.

[97] P.D. Katsabanis, M. Rizk, "Underwater Performance of Polyurethane – PETN Explosive". *Journal of Energetic Materials*, Vol. 10, No. 2-3, 1992, pp. 97–126.

M = 22.47013939 g/mol

Q = 1276.42283 cal/g

phi = 5.141275512

$D_{0.6\ g/cc} = 1.01\ (1 + 1.3 \times 0.6)\ \sqrt{5.141275512} = 4.0764$ km/s.

Kamlet's method is adopted for prediction of detonation velocity and other detonation parameters by various researchers. Nitrobenzenes are evaluated for their heat of formation and change in Gibbs free energy and detonation parameters are estimated by Kamlet's method in paper by Xin *et al.*[98]. The method proposed by Kamlet has been indicated as equation 1 and 2 in another paper for estimation of detonation parameters[99]. Kamlet formulation is used for prediction of detonation parameters for nitro derivative of Pyridines[100].

Approach by Rothstein

In fact Kamlet's empirical relation is not, *the only,* scheme to predict explosive parameters. One of the schemes for prediction of explosive parameters mainly velocity of detonation is given by Rothstein[101]. The scheme is based on molecular structure but it can be extended to explosive compositions, also. However, it is proposed for not only C-H-N-O based explosives but also other elements like chlorine and fluorine[102]. Rothstein *et al.* proposed a parameter 'F' for such calculations, which is a function of many

---

98 Xin Liao, Xue-Hai JU, Feng-Qi Zhao, Jian-Hua Yi, "Substituent Effects on Thermodynamic and Detonation Properties of Polynitrobenzenes". *Propellants Explosives Pyrotechnics*, Vol. 35, No. 6, 2010, pp. 567–571.

99 Peter Politzer, Pat Lane, Jane S. Murray, "Computational Characetrization of a Potential Energetic Compound: 1,3,5,7-Tetranitro-2,4,6,8,-Tetraazacubane". *Central European Journal of Energetic Materials*, Vol. 8, No. 1, 2011, pp. 39–42.

100 Lemi Turker, Selcuk Gumus, Taner Atalar, "A DFT Study on Nitro Derivatives of Pyridine". *Journal of Energetic Materials*, Vol. 28, No. 2, 1010, pp. 139–171.

101 Rothstein L.R., Petersen R, "Predicting High Explosive Detonation Velocities from Their Composition and Structure". *Propellants Explosives Pyrotechnics*, Vol. 4, No. 1, 1979, pp. 56–60.

102 L.R. Rothstein, "Predicting High explosive Detonation Velocities from Their Composition and Structure". *Propellants Explosives Pyrotechnics*, Vol. 6, No. 4, 1981, pp. 91–93.

structural parameters of the explosive molecules like physical state (liquid or solid), molecular structure (aromatic or not), Bonding of oxygen with carbon, Nitrato groups etc. velocity of detonation (D) can be calculated from this factor by a simple linear relation. It is also proposed that calculated velocity of detonation is at the theoretical maximum density (TMD) of the explosive. Density dependent variation of velocity of detonation is not proposed in the correlation. The complete scheme for calculation of VOD is given below.

$$F = [100 \times [n(O) + n(N) - [\{n(H) - n(HF)\}/\{2 \times n(O)\}] + (A/3) - n(B/F)/1.75 - n(C)/2.5 - n(D)/4 - n(E)/5]/MW\ ] - G$$

$$D\ (mm/\mu s) = (F - 0.26)/0.55$$

Where, n(O) = number of oxygen atom (d in earlier expressions), n(N) = number of nitrogen atoms (c in earlier expressions), n(H) = number of hydrogen atoms (b in earlier expressions), n(HF) = number of hydrogen fluoride molecule that will possibly form from available hydrogen, A = 1 if compound is aromatic else zero, n(B / F) = number of oxygen atoms in excess of those already available to form CO_2 *and* H_2O *and / or the number of fluorine atoms in excess of those available to form HF, n(C) = number of oxygen atoms doubly bonded directly to carbon as in ketone or ester, n(D) = number of oxygen atoms singly bonded directly to carbon as in >C-O-R where R is –H,* $-NH_4$*, –C, n(E) = number of nitrato groups existing either in a nitrate ester configuration or as a nitric acid salts, MW = molecular weight of the explosive, G = 0.4 for liquid explosives and zero for solid explosives, D = velocity of detonation.*

The calculation is carried out for RDX for which n(O) = 6, n(N) = 6, n(H) = 6, A = 0, n(B) = 0, n(C) = 0, n(D) = 0, n(E) = 0, MW = 222.174, G = 0.

$$F = [100 \times [6 + 6 - 6/[2 \times 6] + (0/3) - 0/1.75 - 0/2.5 - 0/4 - 0/5]/222.174] - 0 = 5.1761$$

$$D\ (mm/\mu s) = (F - 0.26) / 0.55 = 8.9384$$

Since calculation does not have any term depicting density, the predicted velocity of detonation (D) is always at theoretical maximum density (TMD).

For RDX, TMD is proposed in the report[103] as 1.83 g/cc. This can be compared with velocity of detonation (D) calculated by method proposed by Kamlet. At a density of 1.805 g/cc, VOD of RDX is proposed by Kamlet's formulation as 8.8013 mmμs. Velocity of detonation of RDX at density of 1.83 g/cc is given as $8.8013 \times (1 + 1.3 \times 1.83) / (1 + 1.3 \times 1.805)$ mm/s = 8.8868 mm/s, which is lower than VOD proposed by Rothstein *et al.* The experimental value of VOD is given in the report as 8.85 mm/s.

The calculation is repeated for TNT for which n(O) = 6, n(N) = 3, n(H) = 5, A = 1, n(B) = 0, n(C) = 0, n(D) = 0, n(E) = 0, MW = 227.158, G = 0.

$$F = [100 \times [6 + 3 - 5/[2 \times 6] + 1/3 - 0/1.75 - 0/2.5 - 0/4 - 0/5] / 227.158] - 0 = 3.9253$$

$$D\,(\text{mm}/\mu\text{s}) = (F - 0.26)/0.55 = 6.6642.$$

The experimental value given in the paper is 6.96 mm/μs. As per earlier approach proposed by Kamlet, VOD of TNT at a density of 1.654 g/cc is 7.016 mm/μs, which is nearer to experimental values.

The approach is extended to explosive formulations also, however, giving structural parameters like number of nitrato groups or number of oxygen doubly or singly binder to oxygen etc are difficult to feed accurately for explosive formulations. Rothstein has given scheme for prediction of detonation pressure also. This is obtained by regression analysis of randomly selected 11 explosives. The selection of explosives is such that these represent structure, composition and physical forms of all varieties adequately. The developed empirical relation is given below:

$$P\,(\text{kbar}) = 93.3\,D(\text{km/s}) - 456$$

---

103 L.R. Rothstein, R. Petersen, "Predicting High Explosives Detonation Velocity of from Their Composition and Structure". ADA 062265, NWSY TR 78-4, Naval Weapon Station, Verginia, Dec 1978.

Contrary to Kamlet's approach where pressure is assumed as proportional to square of detonation velocity, this approach proposes a liner variation of detonation pressure with velocity of detonation. This approach is although empirical, but is free from various parameters like density and heat terms. Although this makes calculation simpler dependence of detonation velocity on density cannot be ruled out for high explosives. To take care of this an empirical relation is again proposed in the paper.

> *$D = D_{TMD} - 3 \times (\rho_{TMD} - \rho)$, where D = velocity of detonation at density ρ and D_{TMD} = velocity of detonation at theoretical maximum density as predicted by the Rothstein method from factor 'F'.*

All the governing factors for calculation of F and D for the explosives enlisted in Table 1 are given in Table 10. Last two columns, containing density (TMD in g/cc) and experimental velocity of detonation (D in mm/µs), are taken from the report by Rothstein *et al.* The close matching is observed for all the explosives. Last explosive molecule namely NTO was not experimented at the time of publication of report by Rothstein. Hence, data are not available for this molecule.

For calculation of detonation parameters for explosive compositions, one mole of the mixture must be considered. For a 50:50 RDX/TNT composition, equivalent composition is $C_{4.98}H_{5.505}N_{4.517}O_6$, with a molecular weight of 224.638 g/mol and theoretical maximum density of 1.7312 g/cc. Number of moles of RDX is 0.225, while that for TNT is 0.220. Total number of moles is 0.445. The experimental velocity of detonation is 7.66 km/s for this composition. Out of the two molecules, only TNT is aromatic, which affects parameter A.

A = weight fraction of TNT × mole fraction of TNT
= (50/100) × (0.22/0.445) = 0.247227

F = (100/224.638) × (6 + 4.517 – [5.505/(2 × 6)] + 0.247227/3)
= 4.514

D = (4.514 – 0.26)/0.55 = 7.734 km/s.

P = 93.3 × 7.734 – 456 kbar = 265.64 kbar.

Table 10: Calculation of VOD by method proposed by Rothstein

Explosive	*n(O)*	*n(N)*	*n(H)*	*A*	*B*	*C*	*D*	*E*	*G*	*MW*	*F*	*D*	*VOD Exptl*	*TMD*
RDX	6	6	6	0	0	0	0	0	0	222.17	5.1761	8.9384	8.85	1.83
TNT	6	3	5	1	0	0	0	0	0	227.16	3.9253	6.6642	6.96	1.65
HMX	8	8	8	0	0	0	0	0	0	296.23	5.2324	9.0407	9.14	1.9
PETN	12	4	8	0	0	0	4	0	0	316.18	4.6387	7.9613	8.29	1.77
Tetryl	8	5	5	1	0	0	0	0	0	287.19	4.5339	7.7707	7.91	1.73
HNS	12	6	6	1	0	0	0	0	0	450.28	4.016	6.8291	7.12	1.74
NQ	2	4	4	0	0	0	0	0	0	104.11	4.8028	8.2596	8.16	1.72
Explosive D	7	4	6	1	0	0	1	0	0	246.17	4.3282	7.3967	7.36	1.72
TATB	6	6	6	1	0	0	0	0	0	258.2	4.5829	7.8599	7.94	1.94
DATB	6	5	5	1	0	0	0	0	0	243.18	4.4891	7.6893	7.67	1.84
NTO	3	4	2	0	0	1	0	1	0	130.1	4.6631	8.0056	–	–

The calculated value of velocity of detonation is slightly on higher side than experimentally observed values. However percentage error in calculation is very small. The above-mentioned scheme for prediction of velocity of detonation is used for developing LOTUSES (Linear Output Thermodynamics Userfriendly Software for Energetic Systems) program[104,105].

Method by Xiong

Xiong[106] made an attempt to prediction of detonation velocity of explosives on the basis of almost similar approach as that of Kamlet. He also developed empirical relations on the basis of product of explosion, but the final expressions for calculation of detonation velocity were different. The basic input parameters were density and heat of formation of explosive along with molecular formula of the explosive. Two intermediate parameters namely heat of formation and potential energy are defined and further calculations are based on empirical relation given below.

$D = 67.6 \times \sqrt{Q} + 243.2 \times w \times \rho$, where Q = heat of detonation in cal/g, w = potential energy and ρ = density of the explosive. The numerical constants are derived from the experimental data.

The parameter Q has similar meaning as that in the formulation of Kamlet, but rather than making choice of product of explosion arbitrary, the paper proposed five set of product of explosions depending on oxygen availability in the composition. The formulation of term Q is same as that derived in previous section. The potential energy parameter uses covolumes of product of

104 H. Muthurajan, R. Sivabalan, M.B. Talwar, S.N. Asthana, "Computer Simulation for Prediction of Performance and Thermodynamic Parameters of High Energy Materials". *Journal of Hazardous Materials*, A112, 2004, pp. 17–33.

105 H. Muthurajan, R. Sivabalan, M.B. Talwar, S. Venugopalan, B.R. Gandhe, "Computer Code for the Optimization of Performance Parameters of Mixed Explosive Formulations". *Journal of Hazardous Materials*, A136, 2006, pp. 475–481.

106 Wu Xiong, "A Simple Method for Calculating Detonation Parameters of Explosives". *Journal of Energetic Materials*, Vol. 3, 1983, pp. 263–277.

explosions. This is equal to covolume (k) of product of explosion weighed by moles of product of explosion and divided by molecular weight of the explosive. The calculation of gamma is also proposed in a new way. This relation utilizes gamma values for product of explosion. The calculation of γ_0 for the product of explosion follows harmonic variation and has same mathematical expression as used in calculation of density of a mixture of explosive molecules. Overall, this formulation desire certain input values for product of explosion, which are tabulated below (Table 11).

Potential energy, $w = (\Sigma_{\text{explosion products}}\ n \times k)/MW$

$\gamma = 1.25 + \gamma_0 \times (1 - e^{-0.546 \times \rho})$

Table 11: Input parameters for Xiong's method

Species	*Covolume*	*Heat of formation (cal/mol)*	*Gamma*
H_2O	250	– 57.79	1.68
CO_2	600	– 93.98	3.1
CO	390	– 26.46	2.67
N_2	380	0	3.8
H_2	214	0	3.4
O_2	350	0	3.35
CH_4	528	– 17.80	2.93
C(s)	46	10	3.5

Since all calculations are based on product of explosion, the scheme proposed by Xiong divides all the explosive molecules and formulations into five categories from oxygen rich to serious deficient and for each class; there is a governing constraint. The constraint and product of explosion are tabulated for an explosive molecule of type $C_aH_bN_cO_d$ in Table 12.

The calculation strategy starts with checking, type of molecule or formulation from the mentioned five categories. The checking of category should start from serious deficient (type V) side. Once category is known, product of explosion can be taken from Table 12. After that value of Q and w are calculated from data given in Table 11. Then using relation proposed earlier, velocity of detonation for any explosive can be calculated. The value of gamma can also be calculated and from these two values, detonation pressure can be calculated.

Table 12: Xiong's criteria for product of explosion

Species	***$d-b/2-2a \geq 0$***	***$d-b/2-2a < 0$***	***$d-b/2-a \leq 0$***	***$d-b/2 \leq 0$, $a \geq d$***	***$d-b/2 \leq 0$, $a < d$***
Type	***I***	***II***	***III***	***IV***	***V***
Characteristic	***Rich***	***Slightly deficient***	***Deficient***	***Seriously deficient***	
H_2O	0.5b	0.43b	0.35b	0.54d	d-a
CO_2	a	(d/2-b/4)	(d/2-b/4)	–	–
CO	–	0.07b	0.15b	0.46d	a
N_2	c/2	c/2	c/2	c/2	c/2
H_2	–	–	–	(b/2-d)	(b/2-d+a)
O_2	(d–b/2–2a)/2	–	–	–	–
CH_4	–	0.035b	3b/40	0.23d	–
C(s)	–	(a-d/2+0.145b)	(a-d/2+b/40)	(a-0.69d)	–

The calculation is carried out for RDX $C_3H_6N_6O_6$. Here, a = 3, b = 6, c = 6 and d = 6. Molecular weight of RDX is 222.174. Check for V, d – b/2 = 6 – (6/2) = 3. It is not serious deficient. No need to check for type IV, as first condition is same for the both. Check for III, d – b/2 – a = 6 – (6/2) – 3 = 0, it is deficient. No need to check further. RDX is of type III and the governing equation is given below.

$$C_3H_6N_6O_6 \rightarrow 3N_2 + 2.1\ H_2O + 1.5\ CO_2 + 0.9\ CO + 0.45\ CH_4 + 0.15\ C.$$

$$Q = (14.71 + 2.1 \times 57.79 + 1.5 \times 93.98 + 0.9 \times 26.46 + 0.45 \times 17.80 - 0.15 \times 10)/222.174 = 1383.43 \text{ cal/g}$$

$w = (3 \times 380 + 2.1 \times 250 + 1.5 \times 600 + 0.9 \times 390 + 0.45 \times 528 + 0.15 \times 46)/222.174 = 14.225.$

$D = 67.6 \times \sqrt{Q} + 243.2 \times w \times \rho$

$= 67.6 \times \sqrt{1383.43} + 243.2 \times 14.225 \times 1.805$

$= 8758.78$ m/s

$\gamma_0 = (3 + 2.1 + 1.5 + 0.9 + 0.45 + 0.15)/(3/3.8 + 2.1/1.68 + 1.5/3.1 + 0.9/2.67 + 0.45/2.93 + 0.15/3.5) = 2.6497$

$\gamma = 1.25 + 2.6497 \times (1 - e^{-0.546 \times 1.805}) = 2.91$

Once, velocity of detonation (VOD) and gamma for the explosive, are available, value of detonation pressure can be calculated.

$$P = \rho \times D^2 \times 10^{-5} / (\gamma + 1)$$

$$= 1.805 \times 8758.78^2 \times 10^{-5}/(2.91 + 1) \text{ kbar}$$

$$= 354.15 \text{ kbar}$$

If this formulation is compared with method proposed by Kamlet, complete oxidation of carbon or hydrogen is considered to carbon dioxide monoxide or water and then further calculations are made. Here five sets of conditions are given and accordingly a better estimate of velocity of detonation is possible.

The calculation is repeated for TNT also. TNT also falls in category III of Table 12.

$C_7H_5N_3O_6 \rightarrow 1.5\ N_2 + 1.75\ H_2O + 1.75\ CO_2 + 0.75\ CO + 0.375\ CH_4 + 4.125\ C.$

$Q = (-15 + 1.75 \times 57.79 + 1.75 \times 93.98 + 0.75 \times 26.46 + 0.375 \times 17.80 - 4.125 \times 10)/227.158 = 1038.34$ cal/g

$w = (1.5 \times 380 + 1.75 \times 250 + 1.75 \times 600 + 0.75 \times 390 + 0.375 \times 528 + 4.125 \times 46)/227.158 = 12.0522.$

$D = 67.6 \times \sqrt{Q} + 243.2 \times w \times \rho$

$= 67.6 \times \sqrt{1038.34} + 243.2 \times 12.0522 \times 1.654 = 7026.32$ m/s

$\gamma_0 = (1.5 + 1.75 + 1.75 + 0.75 + 0.375 + 4.125)/(1.5/3.8 + 1.75/1.68 + 1.75/3.1 + 0.75/2.67 + 0.375/2.93 + 4.125/3.5) = 2.8564$

$\gamma = 1.25 + 2.8564 \times (1 - e^{-0.546 \times 1.654}) = 2.9487.$

$P = \rho \times D^2 \times 10^{-5} / (\gamma + 1)$

$= 1.654 \times 7026.32^2 \times 10^{-5} / (2.9487 + 1)$ kbar $= 206.8$ kbar

Table 13: Calculation of salient parameters by Xiong's method

Explosive	*Type*	H_2O	CO_2	*CO*	CH_4	O_2	N_2	H_2	*C*	*MW*	*Q*	*w*	*D*	*Gamma*	*P det*
RDX	3	2.1	1.5	0.9	0.45	0	3	0	0.15	222.17	1383.4	14.225	8762	2.9113	354.52
TNT	3	1.75	1.75	0.75	0.375	0	1.5	0	4.125	227.16	1038.3	12.052	7026	2.9487	206.79
HMX	3	2.8	2	1.2	0.6	0	4	0	0.2	296.23	1377.8	14.225	9100	2.9633	398.01
PETN	2	3.44	4	0.56	0.28	0	2	0	0.16	316.18	1468.2	13.896	8606	2.7896	347.86
Tetryl	3	1.75	2.75	0.75	0.375	0	2.5	0	3.125	287.19	1251.8	12.785	7771	3.0182	259.99
HNS	3	2.1	4.5	0.9	0.45	0	3	0	8.15	450.28	1139.9	11.834	7290	3.1063	225.19
NQ	5	1	0	1	0	0	2	1	0	104.11	582.58	15.503	8456	3.0034	323.29
Explosive D	3	2.1	2	0.9	0.45	0	2	0	2.65	246.17	896.3	12.981	7444	2.9246	242.45
TATB	3	2.1	1.5	0.9	0.45	0	3	0	3.15	258.2	874.51	12.775	8020	3.1016	303.92
DATB	3	1.75	1.75	0.75	0.375	0	2.5	0	3.125	243.18	975.69	12.632	7755	3.0702	271.42
NTO	3	0.7	1	0.3	0.15	0	2	0	0.55	130.1	962.65	13.501	8435	3.18	328.47

The calculations are repeated for each of the molecules mentioned in Table 1 and the intermediate parameters using method proposed by Xiong are tabulated as Table 13. Here detonation velocity is indicated in m/s while detonation pressure (last column) is expressed in kbar.

Although the calculation is mainly concentrating on explosive molecules but calculations can be carried out easily for those compositions, where ingredients can be expressed in terms of C-H-N-O. Let us take for example composition 1 of Table 6, which is made of RDX/TNT in the ratio 75:25. Density of the mixture is calculated as 1.75 g/cc and molecular weight is calculated earlier as 223.23 g/mol. Since all parameters for both RDX and TNT are available, heat of detonation of the mixture is weighed sum of individual heat of detonations. It is the case with potential energy also. However variation of gamma for the mixture follows harmonic summation.

$$Q_{mixture} = 0.75 \times 1383.43 + 0.25 \times 1038.34 \text{ cal/g}$$
$$= 1297.16 \text{ cal/g.}$$
$$w_{mixture} = 0.75 \times 14.225 + 0.25 \times 12.052 = 13.682$$
$$D = 67.6 \times \sqrt{Q} + 243.2 \times w \times r$$
$$= 67.6 \times \sqrt{1297.16} + 243.2 \times 13.682 \times 1.75$$
$$= 8257.75 \text{ m/s}$$
$$1/(223.23 \times \gamma) = [0.75/(222.174 \times 2.9113)] + [0.25/(227.158 \times 2.9487)]$$
$$\text{So,} \quad \gamma = 2.9226$$
$$P = \rho \times D^2 \times 10^{-5} / (\gamma + 1)$$
$$= 1.75 \times 8257.75^2 \times 10^{-5} / (2.9226 + 1) \text{ kbar}$$
$$= 304.22 \text{ kbar}$$

Formulations containing other elements like fluorine, chlorine and aluminium can also evaluated by this formulation, if values of heats of formation are known. For example, for aluminium, value of Q can be taken as 3710 cal/g, value of potential energy is –1 and gamma can be taken as 4. Molecular weight of Al is 26.98. Using this information, Aluminized compositions can be tackled by this method comfortably. Second composition of Table 6 can be taken as trial case for this calculation. It has RDX/TNT/Al in the ratio 42:40:18. The molecular weight and density are calculated as 96.89 g/mol and 1.83 g/cc.

$$Q_{mixture} = 0.42 \times 1383.43 + 0.40 \times 1038.34 + 0.18 \times 3710 \text{ cal/g} = 1664.176 \text{ cal/g.}$$

$$w_{mixture} = 0.42 \times 14.225 + 0.40 \times 12.052 + 0.18 \times -1 = 10.615$$

$$D = 67.6 \times \sqrt{Q} + 243.2 \times w \times \rho$$

$$= 67.6 \times \sqrt{1664.176} + 243.2 \times 10.615 \times 1.83$$

$$= 7482.1 \text{ m/s}$$

$$1/(96.89 \times \gamma) = [0.42/(222.174 \times 2.9113)] + [0.40/(227.158 \times 2.9487)] + [0.18/(26.98 \times 4)]$$

So, $\gamma = 3.541$

$$P = \rho \times D^2 \times 10^{-5} / (\gamma + 1)$$

$$= 1.83 \times 7482.1^2 \times 10^{-5}/(3.541 + 1) \text{ kbar} = 225.6 \text{ kbar}$$

The calculation can be extended to include other ingredients in the explosive formulations, provided 'Q' , 'w' and 'γ_0' are known for these ingredients. Another approach for prediction of detonation parameter is proposed by Hel Karel[107]. He emphasized use of bulk velocity of sound in the explosives and used some of the variables calculated as per method proposed by Xiong to arrive at certain empirical relations for VOD. For a CHNO explosive, oxygen factor (α) is given by [2d/(4a + b)]. Since entire heat release (Q) does not take part in detonation, a release factor is conceived, which is dependent on density and oxygen balance. The correlation is release factor (η) = (α + 1.65) × ρ / 5.5. The bulk velocity of sound, $c = (\rho \times \Sigma_i\{z \times \beta\}/M)^3$, where z is number of chemical bond and β is contribution of individual bonds, M is molecular weight. Contributions of all the bonds are added.

Particle velocity, $u_j = [2(\gamma - 1)\,\eta Q/(\gamma + 1)]^{1/2}$.

$$D = [(\gamma + 2)\, c/(\gamma + 1)] + [2(\gamma + 1)/\gamma] \times \sqrt{[2(\gamma - 1)\,\eta Q/(\gamma + 1)]}$$

The calculation is depicted in the paper for RDX. Input

107 Hel Karel, "Comments on "A Simple Method for Calculating Detonation Parameters of Explosives". *Journal of Energetic Materials*, Vol. 10, No. 4-5, 1992, pp. 287–298.

parameters considered from earlier data and Xiong's paper, are as follows for RDX.

$$\rho = 1.806 \text{ g/cc}$$
$$M = 222.174 \text{ g/mol}$$
$$Q = 1383.43 \text{ cal/g} = 5792.4 \text{J/g}$$
$$\gamma = 2.91$$

It has 6 C-H bonds, 6 C-N bonds and 3 N-NO_2 bonds. The contribution of these bonds in the calculation of sonic velocity is given below.

$$\Sigma_i\{z \times \beta\} = 6 \times 95.2 + 6 \times 20.7 + 3 \times 330 = 1685.4$$
$$c = (\rho \times \Sigma_i\{z \times b\}/M)^3 = [1.806 \times 1685.4/222.174]^3$$
$$= 2571.5 \text{ m/s}$$
$$\alpha = 2d/(4a + b) = (2 \times 6)/(4 \times 3 + 6) = 2/3 = 0.667$$
$$\eta = (\alpha + 1.65) \times \rho/5.5 = (0.667 + 1.65) \times 1.806 / 5.5$$
$$= 0.76071$$
$$\text{Particle velocity, } u_j = [2(\gamma - 1)\, \eta Q / (\gamma + 1)]^{1/2}$$
$$= [2 \times (2.91 - 1) \times 0.76071 \times 5792.4 \times 10^3/(2.91 + 1)]^{1/2}$$
$$= 2074.8 \text{ m/s}$$
$$D = [(\gamma + 2)c/(g + 1)] + [2(\gamma + 1)/\gamma] \times \sqrt{[2(\gamma - 1)\eta Q/(\gamma + 1)]}$$
$$= [(\gamma + 2)c/(\gamma + 1)] + [2(\gamma + 1)/\gamma] \times u_j$$
$$= [(2.91 + 2) \times 2571.5/(2.91 + 1)] +$$
$$[2 \times (2.91 + 1)/2.91] \times 2074.8 = 8804.75 \text{ m/s}$$

The method is based on calculation of gamma (γ) accurately, as it is the main control parameter. The calculation is also extended for the mixture of explosives. The calculation of equivalent density, gamma etc remains same but calculation of sonic velocity is modified. For this value of sonic velocities of individual molecules is considered.

$$c_{mixture} = 1/\,[\rho_{mixture}\, \Sigma(X_i/\rho_i c_i)]$$

For example for RDX/TNT 50/50, following inputs are considered.

Equivalent formula: $C_{4.9975}H_{5.4705}N_{4.4818}O_{5.9549}$

$\rho_{RDX} = 1.806$ g/cc, $c_{RDX} = 2571.4$ m/s

$\rho_{TNT} = 1.654$ g/cc, $c_{TNT} = 2162.0$ m/s

M = 223.4 g/mol

Q = 5062 J/g

$\gamma = 2.91$

$\rho_{mixtutre} = 100/[(50/1.806) + (50/1.654)]$ g/cc = 1.75 g/cc

$c_{mixture} = 1/[\rho_{mixture}\, \Sigma(X_i/\rho_i c_i)] = 1/[1.75 \times \{(0.5/1.806 \times 2571.4) + (0.5/1.654 \times 2162)\}] = 2308.89$ m/s

$\alpha = 2d/(4a + b) = (2 \times 5.5949)/(4 \times 4.9975 + 5.4705) = 0.4395$

$\eta = (\alpha + 1.65) \times \rho/5.5 = (0.4395 + 1.65) \times 1.75/5.5 = 0.66484$

Particle velocity, $u_j = [2(\gamma - 1)\, \eta Q/(\gamma + 1)]^{1/2}$

$= [2 \times (2.91 - 1) \times 0.66484 \times 5062 \times 10^3/(2.91 + 1)]^{1/2}$

= 1813.27 m/s

$D = [(\gamma + 2)\, c/(\gamma + 1)] + [2(\gamma + 1)/\gamma] \times \sqrt{[2(\gamma - 1)\eta Q/(\gamma + 1)]}$

$= [(\gamma + 2)\, c/(\gamma + 1)] + [2(\gamma + 1)/\gamma] \times u_j$

$= [(2.91 + 2) \times 2308.89/(2.91 + 1)] + [2 \times (2.91 + 1)/2.91] \times 1813.27 = 7772.17$ m/s

This method requires determination of sonic velocity, which is difficult if database for contribution of bonds to sonic velocity is not known. This method requires inputs such as heat of detonation, which require certain input parameters regarding product of detonation. This again leads to arbitrariness in the calculation and method comes under empirical approach for detonation parameter prediction.

Other Methods for VOD Calculation

Another method for prediction of detonation parameter is proposed by Stine[108]. He banked on dependence of properties of explosives on their molecular structure. The paper argued that RDX and HMX pressed to same density give same detonation velocity, so rather

[108] J.R. Stine, "On Predicting Properties of Explosives – Detonation Velocity". *Journal of Energetic Materials*, Vol. 8, No. 1-2, 1990, pp. 41–73.

than concentrating on molecular formula, a normalized formula of explosive may be considered. The validity of formulation of C-H-N-O based explosives is also highlighted in the paper. These four elements are placed at the vertex of a tetrahedron with each side unity. The normalized molecular formula allows various explosives to be placed inside this tetrahedron. The TATB can be placed at the center of the virtual tetrahedron. Since detonation velocity depends on both density and heat of formation, normalization desires rationalization of these parameters also. Arbitrarily, 1.8 g/cc and 0 cal/mole is considered as normalized value of density and heat of formation respectively. Then using data of known explosives and predictions by RUBY code, a correlation is developed by regression analysis for prediction of velocity of detonation.

D (km/s) = $3.69 + \rho\,(-13.85a + 3.95b + 37.74c + 68.11d + 0.6917\Delta H_f)/MW$, where ρ = density of the molecule in g/cc, Molecular formula of the explosive is $C_aH_bN_cO_d$, MW = molecular weight of the explosive and ΔH_f = heat of formation of the explosive in kcal/mol.

Table 14: VOD prediction by Stine

Explosive	*VOD(m/s)*
RDX	8790.19
TNT	6852.29
HMX	9062.28
PETN	8428.04
Tetryl	7663.42
HNS	7115.62
NQ	8433.04
Explosive D	7200.49
TATB	7819.71
DATB	7600.64
NTO	8520.17

All the explosives given in Table 1 are subjected to calculation as per this simple method proposed by Stine and results are tabulated in Table 14. The formulation is simple, straightforward and with minimum input parameters.

Keshavarz have proposed several relations for prediction of detonation parameters of explosives, but his methods are based on regression analysis. One of the significant relations for prediction of VOD is based on estimation of approximate detonation temperature[109]. For this first of all, difference in heat of formation (in kcal/mol) of detonation product and that of explosive is calculated. Then detonation temperature is calculated using the following relations. Detonation velocity is calculated subsequently.

$$T_{det} = [298.15\ \Sigma C_{p,i} + \Delta H_{f,explosive} - \Delta H_{f,\ products}]/\Sigma\ C_{p,i}$$

Where, $\Sigma C_{p,i}$ is sum of heat capacities (in kcal/mol/K) of all the I-th components of the detonation products at or near detonation temperature.

$$D\ (km/s) = 0.314 \times \rho \times \sqrt{(N \times T_{det})} + 1.95$$

Where, ρ = density of the explosive, N = number of gaseous molecules per unit weight of explosive (as calculated in Kamlet's formulation).

One of the biggest problems with this method is estimation of specific heat of different product of explosions at detonation temperatures. To take care of this specific heat is assumed to be polynomial function of temperature for each of the product gases. The relation is like $C_{p,i} = a + bT + cT^2 + dT^3 + eT^4 + fT^5$. For the product gases the relations are given below with specific heat calculated in kJ/kg/K.

For CO_2: $C_{p,i} = 0.818205 + 9.9739 \times 10^{-4}\,T - 7.61047 \times 10^{-7}\,T^2 + 2.79744 \times 10^{-10}\,T^3 - 3.8726 \times 10^{-14}\,T^4$.

For CO: $C_{p,i} = 1.030 + 0.1274 \times 10^{-3}\,T + 0.2414 \times 10^{-6}\,T^2 - 0.2174 \times 10^{-9}\,T^3 + 0.4956 \times 10^{-13}\,T^4$.

For H_2O: $C_{p,i} = 1.86024 + 3.23229 \times 10^{-4}\,T + 5.84858 \times 10^{-7}\,T^2 - 3.5846 \times 10^{-10}\,T^3 + 5.93307 \times 10^{-14}\,T^4$.

For O_2: $C_{p,i} = 0.9057 + 0.2941 \times 10^{-3}\,T + 0.9650 \times 10^{-7}\,T^2 - 0.3364 \times 10^{-9}\,T^3 + 0.2021 \times 10^{-12}\,T^4 - 0.3811 \times 10^{-16}\,T^5$.

---

[109] M.H. Keshavarz, H.R. Pouretedal, "Predicting the Detonation Velocity of CHNO Explosives by a Simple Method". *Propellants Explosives Pyrotechnics*, Vol. 30, No. 2, 2005, pp. 105–108.

For N_2: $C_{p,i} = 1.036709 + 8.11879 \times 10^{-6}T + 4.859 \times 10^{-7}T^2 - 4.61612 \times 10^{-10}T^3 + 1.6814 \times 10^{-13}T^4 - 2.18231 \times 10^{-17}T^5$.

For H_2: $C_{p,i} = 14.350 + 0.9947 \times 10^{-3}T - 0.2480 \times 10^{-5}T^2 + 0.4978 \times 10^{-8}T^3 - 0.2856 \times 10^{-11} T^4 + 0.5275 \times 10^{-15} T^5$.

For RDX, the detonation products are given as $3CO + 3N_2 + 3H_2O$, for which value of N is calculated earlier as 0.405087. Keshavarz *et al.*, has used gas phase heat of formation of the explosives from software and molecular modeling rather than using experimental value of heat of formation. The value of detonation temperature for RDX is calculated as 3750 K in the paper. Now detonation velocity, D (km/s) = $0.314 \times \rho \times \sqrt{(N \times T_{det})} + 1.95 = 0.314 \times 1.805 \times \sqrt{(0.0405087 \times 3750)} + 1.95 = 8.9355$ m/s. this value seems to be very high. However, in the paper calculations are made for various explosives including TNT, DATB, Tetryl, PETN, PA, NQ at different loading densities and close matching to BKW-EOS as well as experimental values of VOD is claimed.

In another paper, Keshavarz[110] has proposed a linear relation for prediction of detonation velocity using molecular formula, density and heat of formation. The relation resembles Stine's correlation[108], but has a different format. The relation is given below.

$$D\ (\text{km/s}) = (-3.748/\rho) - 13.438a + 13.033b - 28.63c + 24.704d + 0.7265\,\Delta H_f$$

Where ρ = density of explosive in g/cc, a, b, c, d are number of atoms of carbon, hydrogen, nitrogen and oxygen atoms, ΔH_f is crystalline heat of formation in kcal/mol.

This relation is claimed to be valid for explosives with nonideal detonation behaviour. The new correlation is useful for ammonium nitrate based explosives, where latent heat losses and presence of rarefaction waves extinguishes decomposition reaction giving a wide reaction zone. The behaviour results in situations of no reaction to partial reaction to complete reaction in assorted cases, giving wide variation in performance. The estimation by BKW-

---

110 M.H. Keshavarz, "Prediction of Detonation Velocities of Non-Metal Nitrated Explosives by Simple Method". *Asian Journal of Chemistry*, Vol. 17, No. 4, 2005, pp. 2223–2228.

EOS code fails to predict such variations because of non-application of steady equilibrium one-dimensional theories. The correlation is developed for such explosives formulation where ammonium nitrate is one of the ingredients to impart non-ideal behaviour.

The first example cited is bare ammonium nitrate (NH_4NO_3), whose molecular weight is 80 g/mol. On 100 g basis, the molecular formula can be given as $C_0H_5N_{2.5}O_{3.75}$. The heat of formation and density of the molecule are –109.25 kcal/mol and 0.7 g/cc. For such situations, experimental velocity of detonation is found to be 1.50 km/s. Using relation proposed above, the value of detonation velocity is calculated as 1.43 km/s. For Amatex-20 (RDX/TNT/AN: 20/40/40), the equivalent molecular formula for 100 g molecular weight of the equivalent compound is $C_{1.73}H_{2.99}N_{2.14}O_{2.90}$. At density of 1.66 g/cc and heat of formation of –22.74 kcal/mol, the experimental VOD is 7.55 km/s. By the method proposed above, the calculated VOD is 7.311 km/s. However, the relation may not be very useful for conventional high explosive molecules not containing AN. For RDX, the equivalent molecular formula for a molecular weight of 100 g/mol is $C_{1.35}H_{2.70}N_{2.70}O_{2.70}$. At density of 1.805 g/cc and heat of formation of 14.71 kcal/mol, the calculated VOD using formula referred above is 15.051 km/s, which is nowhere near the predicted value or experimental value of VOD for RDX. A unique surprise comes for PETN, which has an equivalent molecular formula for 100 g/mol molecular weight as $C_{1.58}H_{2.53}N_{1.265}O_{3.79}$. It has a density of 1.78 g/cc and heat of formation of –128.74 kcal/mol as given in Table 1. The calculated value of VOD is –26.343 km/s. Obviously, the formulation may be valid for mixture of explosives containing ammonium nitrate but not suitable for other explosives and formulations, where steady ideal detonation with complete reaction takes place. This method is simple but it has limited acceptability.

For high explosive molecules, Keshavarz[111] has given another simple relation, where heat of formation is not used as an input parameter. The results are found to be matching to that predicted by BKW-code. The relation is given below.

---

[111] M.H. Keshavarz, "Prediction of Detonation Velocities of Non-Metal Nitrated Explosives by Simple Method". *Asian Journal of Chemistry*, Vol. 17, No. 4, 2005, pp. 2223–2228.

D (km/s) = 1.6439 + 3.5933ρ – 0.1326a – 0.0034b + 0.1206c + 0.0442d – 0.2768n_{NRR}, Where ρ = density of explosive in g/cc, a, b, c, d are number of atoms of carbon, hydrogen, nitrogen and oxygen atoms, n_{NRR} is number of specific.

Groups like –NH2, NH4+, 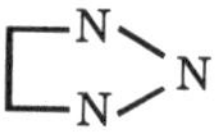

For RDX, there is no special group and the velocity of detonation in km/s is given by (1.6439 + 3.5933 × 1.805 – 0.1326 × 3 – 0.0034 × 6 + 0.1206 × 6 + 0.0442 × 6 – 0.2768 × 0 =) 8.704 km/s. In the paper, reported value of VOD of RDX at density of 1.80 g/cc is given as 8.75 km/s, which becomes 8.77 km/s at density of 1.806 g/cc, used in this book. The calculations are repeated for all the molecules mentioned in Table 1 and VOD is found to be matching to other researchers as well as reported values. Since values of VOD are reported at different densities, highest density is considered and value of VOD is depicted in the table as reported value (in km/s) at certain specified densities. The reported values of VOD are adjusted to density at which VOD of the explosives is calculated. The data are tabulated as Table 15.

$$D_{adjusted} = D_{reported} + 3.5933 \times (\rho_{table1} - \rho_{reported})$$

Table 15: VOD calculated by linear method of Keshavarz

Explosive	*VOD (km/s)*	*Adjusted VOD (km/s)*	*Reported VOD (km/s)*	*at Density (g/cc)*	*Sp Group*
RDX	8.704	8.772	8.75	1.8	0
TNT	7.269	6.980	6.93	1.64	0
HMX	9.250	9.164	9.11	1.89	0
PETN	8.363	8.342	8.27	1.76	0
Tetryl	7.872	7.720	7.72	1.73	0
HNS	7.273	7.303	6.8	1.6	0
NQ	8.295	8.698	8.59	1.78	1
Explosive D	7.512	7.450	6.85	1.55	1
TATB	7.950	7.968	7.76	1.88	3
DATB	7.747	7.733	7.6	1.8	2
NTO	**8.922**	–	–	–	**0**

A brief review of various empirical methods for the calculation of velocity of detonation of explosives was presented in a paper[112]. The Kamlet's method was assumed to be accurate and precise with respect to the practically observed values.

Complexities and accuracy of other schemes were judged. In another semi-empirical approach[113], the explosive formulation is represented by $C_xH_yN_zO_u$. The product of explosion was estimated from coefficient of realization (K), which depended on oxygen coefficient (K_k) by the relation given below.

$$K = 0.32\ K_k^2$$

This coefficient of realization determines the degree of hydrogen oxidation and formation of water. It was assumed that from y/2 moles of hydrogen, Ky/2 moles of water would be formed and remaining hydrogen [(1 – K)y/2] would be utilized for the formation of hydrogen molecule. So the reaction during explosion was written as given below.

$$C_xH_yN_zO_u \rightarrow n_1H_2O + n_2H_2 + n_3CO_2 + n_4CO + n_5O_2 + n_6N_2 + n_7C$$

Here the values of some of the coefficients were known. Here, $n_1 = Ky/2$, $n_2 = (1 - K)y/2$, $n_6 = z/2$. For the other coefficients three conditions were specified.

1. For $K_k > 100\%$, $n_3 = (1.4K - 0.4)x$, $n_4 = x - x_3 = 1.4x(1 - K)$, $n_5 = (x - 2n_3 - n_4 - n_1)/2$, $n_7 = 0$.
2. For $K_k < 100\%$ and $u > (x + y/2)$, $n_3 = 0.7(u - y/2)K - 0.4x$, $n_4 = 1.4x - 0.7(u - y/2)K$, $n_5 = 0$, $n_7 = 0$.
3. For $K_k < 100\%$, $n_3 = 1.16u(K - 0.568) - 0.5n_1$, $n_4 = u - (2n_3 + n_1) = u[1 - 2.32(K - 0.568)]$, $n_5 = 0$, $n_7 = x - (n_3 + n_4)$.

---

[112] Himanshu Shekhar, Studies on Empirical Approaches for Estimation of Detonation Velocity of High Explosives. *Central European Journal of Energetic Materials*, Vol. 9, No. 1, 2012, pp. 39–48.

[113] RADUN JEREMIÆ, JOVICA BOGDANOV, Development of a New Model for the Calculation of the Detonation Parameters of High Explosives. *Journal of the Serbian Chemical Society*, Vol. 77, No. 3, 2012, pp. 371–380.

Using above mentioned correlation, product of explosion (n, number of moles/kg) was obtained. Chemical heat release (Q in kJ/kg) is ascertained from heat of formation of products and reactants. The value of velocity of detonation and detonation pressure was calculated at an initial density (g/cm^3) by relation given below.

$$P \text{ (kbar)} = A\,\rho_0^{2}\, nQ$$
$$D \text{ (m/s)} = B + C\rho_0\sqrt{nQ}$$

By regression analysis, values of constants are obtained as A = 0.00048, B = 2264.1, C = 7.7072.

The method is applied for the prediction of velocity of detonation of RDX with properties given in Table 1. Input parameters are x = 3, y = 6, z = 6, u = 6, ρ_0 = 1.805 g/cm^3,

Oxygen coefficient (K_k) is calculated as ratio of number of moles of available oxygen divided by oxygen needed for the complete oxidation of carbon and hydrogen to carbon dioxide and water.

Oxygen coefficient, $K_k = 6 \times 100/(3 \times 2 + 6/2)\% = 66.66\%$

Coefficient of realization, $K = 0.32 \times K_k^{0.24} = 0.876775$

Since K_k is less than 100% and u = x + y/2, third condition is assumed to be applicable. The value of moles of the products are as follows, $n_1 = Ky/2 = 0.876775 \times 6/2 = 2.6303$; $n_2 = (1 - K)y/2 = (1 - 0.876775) \times 6/2 = 0.36967$; $n_3 = 1.16u(K - 0.568) - 0.5n_1 = 1.16 \times 6 \times (0.876775 - 0.568) - 0.5 \times 2.6303 = 0.83391$; $n_4 = u - (2n_3 + n_1) = 6 - (2 \times 0.83391 + 2.6303) = 1.70185$; $n_5 = 0$, $n_6 = 6/2 = 3$; $n_7 = x - (n_3 + n_4) = 3 - (0.83391 + 1.70185) = 0.46423$. Mass balance for each element is checked.

$$C : n_3 + n_4 + n_7 = 0.83391 + 1.70185 + 0.46423 = 3$$
$$H : 2n_1 + 2n_2 = 2 \times (2.6303 + 0.36967) = 6$$
$$N : 2n_6 = 2 \times 3 = 6$$
$$O = n_1 + 2n_3 + n_4 + 2n_5$$
$$= 2.6303 + 2 \times 0.83391 + 1.70185 + 2 \times 0 = 6$$

Number of moles of gaseous products is calculated as calculated earlier by adding moles of all the product species except carbon (solid not gaseous). Since value is needed in mol/kg, the value of number of moles calculated is multiplied by 1000 and divided by molecular weight (=222.174) of RDX. Chemical heat release is calculated as given in chapter 3 from product species. Using specified formula detonation velocity and pressure are also calculated. Both the values are near to practical results.

Number of moles, n = 38.42 mol/kg

Chemical heat release, Q = 5467.453 kJ/kg

$P\ (\text{kbar}) = 0.00048\ \rho_0^2\ nQ = 328.86$

$D\ (\text{m/s}) = 2264.1 + 7.7072\rho_0\sqrt{nQ} = 8643.52$

The method is repeated for prediction of detonation parameters of TNT also. Input parameters are x = 7, y = 5, z = 3, u = 6, $\rho_0 = 1.654$ g/cm^3.

Oxygen coefficient (K_k) is calculated as ratio of number of moles of available oxygen divided by oxygen needed for the complete oxidation of carbon and hydrogen to carbon dioxide and water.

Oxygen coefficient, $K_k = 6 \times 100/(7 \times 2 + 5/2)\% = 36.36\%$

Coefficient of realization, $K = 0.32 \times K_k^{0.24} = 0.758071$

Since K_k is less than 100% and $u < x + y/2$, third condition is applicable. The value of moles of the products are as follows, $n_1 = Ky/2 = 1.8952$; $n_2 = (1 - K)y/2 = 0.6048$; $n_3 = 1.16u(K - 0.568) - 0.5n_1 = 0.3753$; $n_4 = u - (2n_3 + n_1) = 3.3542$; $n_5 = 0$, $n_6 = 1.5$; $n_7 = x - (n_3 + n_4) = 3.2705$. Mass balance for each element is checked.

$C : n_3 + n_4 + n_7 = 7$

$H : 2n_1 + 2n_2 = 5$

$N : 2n_6 = 3$

$O = n_1 + 2n_3 + n_4 + 2n_5 = 6$

Number of moles of gaseous products is calculated as calculated earlier by adding moles of all the product species except carbon (solid not gaseous). Since value is needed in mol/kg, the value of number of moles calculated is multiplied by 1000 and divided by molecular weight (=227.158) of TNT. Chemical heat release is calculated as given in chapter 3 from product species. Using specified formula detonation velocity and pressure are also calculated. Both the values are near to practical results.

Number of moles, n = 34.027 mol/kg

Chemical heat release, Q = 4028.26 kJ/kg

$P\ (\text{kbar}) = 0.00048\ \rho_0^2\ nQ = 179.99$

$D\ (\text{m/s}) = 2264.1 + 7.7072\rho_0\sqrt{nQ} = 6983.67$

Detonation parameters of all the explosive molecules given in Table 1 are calculated by this method and the results are given in Table 16. All of the calculated values are matching to earlier calculated values and a different approach is proposed in the paper. Although values do not match exactly with those reported in the paper, but value of heat of formation taken in paper may not be the same as in the calculation, shown.

Table 16: Calculation by semi-empirical method

Explosives	*HMX*	*PETN*	*Tetryl*	*HNS*	*NQ*	*Exp D*	*TATB*	*DATB*	*NTO*
VOD (m/s)	8979	8412	7655	7248	7214	6853	7475	7395	7637
P (Kbar)	364	305	235	201	198	170	219	213	233

Controlling Factors for VOD

For pressed TNT charges, Urizar *et al.*[114] experimentally determined dependence of velocity of detonation (VOD) on density. They found linear variation of velocity of detonation with density as D (in m/s) = 1872.7 + 3187.2 × (density in g/cc) for 0.9 g/cc ≤ density ≤ 1.5342 g/cc. However, for higher densities, the variation becomes quadratic. For density between 1.5342 g/cc and 1.636 g/cc, $D = 6762.5 + 3187.2\,(\text{density} - 1.5342) - 25.102 \times (\text{density} - 1.5342)^2 + 115.056 \times (\text{density} - 1.5342)^3$. Donna Price[115] also proposed a

[114] M.J. Urizar, E.James, L.C. Smith, "Detonation Velocity of Pressed TNT". *Physics of Fluid*, Vol. 4, No. 2, 1961, pp. 262–274.

[115] Donna Price, "The Detonation Velocity – Loading Density Relation for Selected Explosives and Mixtures of Explosives". *Journal of Energetic Materials*, Vol. 1, No. 1, 1983, pp. 55–82.

linear variation of detonation velocity (D) with density. The additive laws for calculation of density for high porosity PETN charges were found acceptable. Urizar's method was found to be unacceptable for high porosity PETN mixtures. The VOD density relation for PETN and RDX is reproduced below from another paper[116].

PETN : D = 1854 + 3639 × density

PETN : D = 1966 + 3695 × density

RDX : D = 2427 + 3307 × density

The first relation for PETN is reproduced from experimental observations and second is from modified BKW code for low density PETN. The relation for RDX is also derived from BKW-code and the relation over-predicts VOD at low densities.

However, Kamlet's method was found satisfactory for a wider density ranges. In fact Kamlet's method also takes care of density dependent variation of velocity of detonation and at two different densities, the following relation relates detonation velocities.

D_1/D_2 = (1 + 1.3 × first density)/(1 + 1.3 × second density)

Rothstein's method does not take care of density dependent variation of velocity of detonation. It calculates VOD at TMD and no relation is provided to correlate it with values at different densities. Xiong's method, being similar to Kamlet's method, accounts for density dependence of velocity of detonation. The final expression for VOD has density dependent term and VOD at two different densities for a given explosive are related by the following expression.

D_1/D_2 = (A + B × first density)/(A + B × second density), where $A = 67.6\sqrt{Q}$ and B = 243.2 × w of the explosive.

In this case, for each molecule, separate constants and slope term in the expression exists, while in case of Kamlet's method,

---

[116] P.D. Katsabanis, M. Rizk, "Underwater Performance of Polyurethane – PETN Explosive". *Journal of Energetic Materials*, Vol. 10, No. 2-3, 1992, pp. 97–126.

constant term is 1 and slope is equal to 1.3. Sil'vestrov[117] proposed similar final relation in his paper, with different coefficients for different explosive molecules.

Some of the researchers tried to correlate velocity of detonation with other parameters also. Kock *et al.*[118] proposed a simple relation to correlate detonation velocity (D) with gurney energy ($\sqrt{2E_G}$), as $\sqrt{2E_G} = 3\sqrt{3} \times D/16 \simeq D/3.08$. This is based on adiabatic expansion of gaseous products during detonation as per ideal theory. Microscopic model of detonation in condensed explosives is developed by Pozdnyakov[119]. This uses four parameters characterizing all the explosive molecules namely vibration wave number ν, molecular weight μ, density ρ and anisotropy factor ξ. The value of this factor is 1, if lattice structure of the explosive is isotropic or as hexagonal packing of sphere (HPS). The final relation for estimation of velocity of detonation (D in m/s) is given below.

$$D\ (m/s) = 1.27 \times \xi \times \nu \times (\mu/\rho)^{1/3}$$

For RDX, values of ν, μ and ρ are 1270/cm, 222 g/mol and 1.80 g/cc, respectively. Since RDX is considered as hexagonal packing of spheres (HPS), value of ξ is taken as 1. From the formula given above, detonation velocity is estimated as 8028.5 m/s. Similarly for TNT, values of ξ, ν, μ and ρ are 1.127, 934/cm, 227 g/mol and 1.64 g/cc, respectively. As per proposed method velocity of detonation can be calculated as 6915.2 m/s. Similarly for HMX, values of ξ, ν, μ and ρ are 1.035, 1289/cm, 316 g/mol and 1.90 g/cc, respectively. As per proposed method velocity of detonation can be calculated as 9002.6 m/s. The method is based on non-empirical microscopic model and it gives detonation velocity within 10% of the experimental results. However, the model is valid for high-density state of explosives, detonating in steady

[117] V.V. Sil'vestrov, "Denstity Dependence of Detonation Velocity for Some Explosives". *Combustion, Epxlosion and Shock Waves*, Vol. 42, No. 4, 2006, pp. 472–479.

[118] Andre Koch, Niklaus Arnold, Markus Estermann, "A Simple Relation between the Detonation Velocity of an Explosive and its Gurney Energy". *Propellants Explosives Pyrotechnics*, Vol. 27, No. 6, 2002, pp. 365–368.

[119] A.V. Pozdnyakov, "Calculation of Detonation Velocity". *Combustion Explosion and Shock Waves*, Vol. 38, No. 3, 2002, pp. 337–345.

manner. If density of RDX is changed to 1.70 g/cc, the model predicts a density of 8182.9 m/s, which is higher than the VOD predicted (8028.5 m/s) at higher density of 1.80 g/cc.

For all one-component explosives, velocity of detonation increases with increase in density. It is known that upper limit for the density is crystal density of the molecule and accordingly maximum velocity of detonation for single molecule is realized at crystal density. In early days (1967), analysis for PETN revealed a nonlinear equation of the form D (km/s) = $2.8 \times (1 + \rho^{1.3})$, where ρ is density in g/cc. However, for most of the explosives, variation of velocity of detonation with density is linear in the range of density higher than 1 g/cc. However for multi-component explosives, velocity of detonation first increases with increase in density then reduces. The decrease in velocity of detonation with increase in density occurs in a state frequently referred to dead pressing. In this case, rise in density beyond certain limit for specified explosive formulations, leads to reduced diffusion rate. This reduces augmentation of physical shock by the energy of chemical reaction and velocity of detonation shows a declining trend. Since most of explosive gets converted to products at the C-J plane, rather than initial density, density at C-J state is more important. Using BKW (Becker-Kistiakiwsky-Wilson) equation of state, C-J density is correlated to initial density of the materials and the developed relation for single crystal explosives is $\rho_{C\text{-}J\ state} = 0.14 + 1.26 \times \rho_0$. Kamlet and Jacob suggested another expression for the same.

$$\rho_0 = \rho_{C\text{-}J\ state}/(1.470 - 0.05625 \times \rho_{C\text{-}J\ state})$$

Using momentum equation developed in chapter 2, pressure at C-J plane can be correlated to initial density of the explosives.

$p_1 - p_0 = \rho_0 \times u_1 \times D = \rho_0 \times (1 - \rho_0/\rho_1) \times D^2$, where p is in kbar, ρ is in g/cc, D is in km/s. As $p_1 << p_0$, $p_1 = 10 \times \rho_0 \times [0.140 + 0.260\rho_0]/[0.140 + 1.260\rho_0] \times D^2$. for density lying between 0.8 g/cc to 1.8 g/cc, this pressure can be written as $p_1 = 2.10 \times [0.360 + \rho_0]\ D^2$.

The velocity of detonation is dependent on charge diameter also for smaller charges. As charge diameter is increased, the dependence becomes insignificant. A conventionally popular equation is given below for the same.

D (km/s) = $D_{ideal} \times (1 - a/d)$, where a is a constant, d is charge diameter of the explosive.

Although 'a' is specified as constant, it varies with change of explosive, density and other parameters. As density of explosive (formulation) increases, velocity of detonation increases and value of 'a' reduces. This indirectly means that as density increases, critical diameter reduces for explosives. However, this reduction in critical diameter with density is confined upto very near to crystal density. At very near crystal density, the critical diameter is found to increase abruptly with rise in density. These phenomena can be explained by hot spot theory of initiation. The hot spots are basically centers of initiation in an explosive. As more number of hot spots are present, explosive reaction rate will be higher due to reduced travel time to bridge gap between two such centers. As shock pressure is higher, the rate of reaction at these centers will be higher. This means higher density gives higher velocity of detonation and detonation pressure. Indirectly as lateral boundaries are relatively near at higher densities, sensitivity of the detonation to disturbances (rarefaction waves) will decrease, giving a lower critical diameter and 'a' value. As density approaches crystal density, this distributed pattern of initiation and sustenance of detonation is converted into a more uniform homogeneous reaction throughout the mass relegating hot spot form of initiation to secondary status. As higher density, reduces number of initiation centers, critical diameter increases near crystal densities. However, RDX/TNT or HMX/TNT based charges conserve their initiation centers even near crystal densities and very moderate increase in crystal diameter result near crystal densities for these explosives. For ANFO type explosives, critical diameter increases with increasing density to a maximum value and then decreases. This means for same critical diameter, two densities are possible at the boundary line. This indirectly means that for each critical diameter, a range of density exists between which, the material does not show detonation. Detonation for such explosives occur only when density is maintained outside the density limits for a given critical charge diameter. The critical diameter generally decreases with decrease in grain size, also. The critical diameter and velocity of detonation also depends on temperature.

Another advanced regression correlation relies on molecular and quantum chemical properties of the molecules[120]. It depicts a linear relation between velocity of detonation and density. The linear dependence of VOD on total energy (E) in au (hartree units) corrected to zero point energy (ZPE) is also obtained. The energy used in the empirical relation are all UB3LYP/6-31G (d) type. The variables used in the regression analysis were density and ratio of total energy to molecular weight of the explosives in grams (E/M). The regression analysis gave following correlation.

$$D \text{ (km/s)} = -7.157115 \times (E/M) + 1.262652 \times \rho - 22.739270$$

Here E is a negative quantity and structures with more negative E values are more stable. For estimating values of E, first geometrical optimization is carried out using molecular mechanics (MM2) method. This is followed by semi-empirical PM3 and quantum chemical calculation (restricted Hartree-Frock and Density Function Theory). These activities are carried out to arrive at the most favorable structures of the species presently considered. Vibrational analysis is also done and each final structure is ensured to have at least a local minimum on the potential energy surface. The bond lengths are also searched to locate any bond cleavage during geometrical optimization process. These all calculations needed a dedicated software package.

Explosives are less stable than non-explosive compounds, so they have more negative E-value. As coefficient of term (E/M) is negative in the expression, this indicates, higher VOD for higher negative value of E. As density increases, VOD also increases. The density dependence follows almost same relation as depicted in case of Rothstein's method earlier. Contrary to Kamlet's method, where multiplier of density in correlation for VOD is 1.3, here 1.262652 is used.

The calculations are carried out for TNT, for which E/M is calculated as –3.89601 au/g. Density of TNT is taken as 1.65 g/cc and experimental VOD is found as 7.02 km/s.

---

[120] P.D. Katsabanis, M. Rizk, "Underwater Performance of Polyurethane – PETN Explosive". *Journal of Energetic Materials*, Vol. 10, No. 2-3, 1992, pp. 97–126.

$$D\,(km/s) = -7.157115 \times (-3.89601) + 1.262652 \times 1.65 - 22.739270$$
$$= 7.2283\ km/s$$

For RDX, density is 1.81 g/cc and value of E/M is – 4.04173 au/g. the velocity of detonation is calculated as 8.473 km/s against experimentally observed value of 8.89 km/s.

It is observed that density is a function of (E/M) and density term can be dispensed with in the equation and final correlation is given below.

$$D\,(km/s) = -8.16353 \times (E/M) - 24.52194$$

This correlation has added advantage of independence from density. For synthesized molecules, density values are readily available but for new or proposed molecules, density is to be estimated. The estimation of density is not needed for use of this equation for estimation of detonation parameter.

Calculations are carried out for the same two molecules TNT and RDX.

For TNT, $D\,(km/s) = -8.16353 \times (-3.89601) - 24.52194 = 7.2832\ km/s$

For RDX, $D\,(km/s) = -8.16353 \times (-4.04173) - 24.52194 = 8.4728\ km/s$

It is observed that for HMX, value of E/M is same as that for RDX, however density of HMX is 1.91 g/cc, which is higher than density of RDX. This indicates that second method, where density is not present, gives same VOD for RDX as well as HMX, which is not true. However, the first method, where density plays a role, gives VOD of HMX as 8.599 m/s. However, both the values are less than experimentally observed value of VOD (9.13 km/s).

A theoretical derivation from the above equations can be made about lower bond of (E/M) values. If VOD is made zero, then positive value of d is possible, only when value of (E/M) is less than 3.17 au/ g. even from the simplified relation, the lower value of (E/M) comes out to be (–24.52194/8.16353 =) –3 au/g. This gives an structural constraint for design of novel explosives. The atoms in kind, structure and bonding have to be arranged in such a way that (E/M) ratio should be less than –3.17 au/g.

Calculation of Other Detonation Parameters

Although, most of the papers calculate, velocity of detonation for explosives and then detonation pressure is calculated by standard correlations. For example by Kamlet's method detonation pressure is derived in terms of density and Kamlet parameter, which are intermediate variables for calculation of VOD. Detonation pressure in general is given by product of density and square of VOD divided by (Gamma + 1) of the explosive. However, there are certain papers, where an approach independent of VOD is attempted. Keshavarz[121] modified approach of Kamlet and proposed modified Kamlet relation as given below for the prediction of detonation pressure.

P (kbar) = $15.88\ \rho^2\phi - 11.17$, where density is in g/cc and ϕ is Kamlet parameter.

In the papers, pathways to predict detonation products of explosives are compiled. Explosives containing Fluorine and Chlorine are also considered along with conventional C-H-N-O based systems. Four pathways are proposed for decomposition of explosive with general formula $C_aH_bN_cO_dF_eCl_f$.

Once decomposition products are known, the values of Kamlet parameters can be ascertained, which will be dependent on the pathways now. Further calculation of detonation pressure can be conducted.

For example for RDX, the decomposition pathway is given below.

$$C_3H_6N_6O_6 \rightarrow 3N_2 + 3CO + 3H_2O$$

---

[121] M.H. Keshavarz, H.R. Pouretedal, "An Empirical Method for Predicting Detonation Pressure of CHNOFCl Explosives". *Thermochemica Acta*, Vol. 414, 2004, pp. 203–208.

This is similar to scheme 2 discussed earlier under Kamlet's method. As per this method, N = 0.0405087, M = 24.667, Q = 1204, $\phi = 6.9805$

Kamlet's method: $P = 15.58\ \rho^2\phi = 352.37$ kbar

Keshavarz's method: $P = 15.88\ \rho^2\phi - 11.17 = 347.98$ kbar

The proposed pathway is given below:

If

$d \le a$ then eHF + f HCl + (c/2) N_2 + d CO + (a – d) C(s) + (b – e – f)/2H_2

Else

If

(b – e – f)/2 > (d – a) then eHF + f HCl + (c/2)N_2 + a CO + (d – a) H_2O + ([(b – e – f)/2] – (d – a))H_2

Else

If

$d \le 2a$ + [(b – e – f)/2] then eHF + f HCl + (c/2)N_2 + ([(b – e – f)/2] H_2O + (2a – d + (b – e – f)/2) a CO + ((d a) – [(b – e – f)/2])CO_2

Else

d > 2a + [(b – e – f)/2] then eHF + f HCl + (c/2)N_2 + ([(b – e – f)/2] H_2O + [(2d – b + e + f)/4) – a] CO + aCO_2

Endif

Endif

Endif

The method is similar to that proposed by Kamlet and it is just adjustments of coefficients and introduction of an intercept term in the formulation. Further expansion of domain of Kamlet's equation is attempted by Gill *et al.*[122]. Being researchers in the

[122] R. Gill, L. Asaoka, E. Baroody, "On Underwater Detonation, I. A New Method for Predicting the CJ Detonation Pressure of Explosives". *Journal of Energetic Materials*, Vol. 5, No. 3-4, 1987, pp. 287–307.

area of solid rocket propellants, they selected specific impulse as one of the controlling factors. The paper gives a relation for detonation pressure for plastic bonded explosives. The final relation is given below.

P (kbar) = 44.4 ρ^2 Isp – 21, where Isp is specific impulse in Ns/g and ρ is density in g/cc.

Keshavarz[123] has further proposed a new relation for calculation of detonation pressure, which is linear relation and dependent on molecular formula, density and heat of formation of the explosive molecules. The new relation is based on HMX, RDX, TNT, PETN, TETRYL and DATB as reference explosives and is given below:

$$P\,(\text{kbar}) = -2.6 + (\rho^2/\text{MW}) \times (-1026a + 226b + 1031c + 3150d + 30.7\Delta H_f)$$

For RDX, ρ = 1.80 g/cc, MW = 222.174 g/mol, a = 3, b = 6, c = 6, d = 6, ΔH_f = 14.71 kcal/mol, P (kbar) = 347.02 kbar. Calculations are repeated for all the molecules of Table 1 and results are given in Table 17.

Table 17: Calculation of detonation pressure

Explosive	*P(kbar)*	*P (Exptl)*
RDX	347.02601	341
TNT	183.83522	210
HMX	385.77461	390
PETN	344.63461	337
Tetryl	252.16353	239
HNS	209.54857	NR
NQ	298.79264	NR
Explosive D	218.80966	NR
TATB	276.02498	260
DATB	251.40827	251
NTO	327.65891	NR

[123] M.H. Keshavarz, "Simple Determination of Performance of Explosives without using any Experimental Data". *Journal of Hazardous Materials*, A119, 2005, pp. 25–29.

The experimental values are also quoted from the paper. NR in the Table 17 indicates 'not reported' values in the paper. The power of this method is that it does not demand calculation of any product of explosion and heat of detonation is not to be evaluated. The method is simple handy and straight forward for a simple handheld calculation.

Since heat of detonation becomes one of the important intermediate parameter for the calculation of detonation velocity or detonation pressure, the calculation cannot be left to some arbitrary scheme. The approach for assumed product of detonation and then calculation of heat of detonation cannot be treated as very authentic and precise. To overcome this, researchers have proposed some definite product of explosion from an explosive by considering different pathways of decomposition. Xiong as well as Keshavarz have tried to bring some definiteness in such formulations. However, Keshavarz[124] has further proposed to calculate heat of detonation without considering any product of explosion. He has proposed two correlations, one valid for aromatic and other for non-aromatic explosives. The final correlation is given below.

For non-aromatic explosives

$$Q\ (\text{kJ/kg}) = (58.722a - 55.011b - 21.234c + 250.92d + 4.485\ \Delta H_f)/\text{MW})$$

For aromatic explosives

$$Q\ (\text{kJ/kg}) = (61.781a - 51.317b - 30.656c + 91.446d - 0.2791\ \Delta H_f)/\text{MW})$$

The values are calculated for all the explosives given in Table 1 and are reported below as Table 18 along with reported experimental values in the paper. Although, relation is very simple to utilize, the calculation require gas phase heat of formation of explosives, which is different from condensed phase heat of formation. The second deviation is unit of heat of formation, which is here kJ/mol rather than kcal/mol as used in earlier correlations.

[124] M.H. Keshavarz, "Simple Procedure for Determine Heats of Detonation". *Thermochemica Acta*, Vol. 428, 2005, pp. 95–99.

Although conversion needs 1 kJ = 4.187 kcal. Experimental values of heat of detonation (Q) are reproduced from the paper. The condensed phase heat of formation in kJ/mol is given as last column, which is different from gas phase heat of formation of explosive as reproduced from the paper (second last column).

Table 18: Calculation of heat of detonation empirically

Explosive	*1 if Aromatic*	*Q (kJ/kg)*	*Q (Exptl)*	*Hf (gas) (kJ/mol)*	*Hf(cond) (kJ/mol)*
RDX	0	6.274241	6.322	37.85	61.59077
TNT	1	3.599812	4.564	−4.3	−62.805
HMX	0	6.191778	6.197	45.02	75.07291
PETN	0	6.318879	6.322	−174.3	−538.867
Tetryl	1	3.68446	4.773	9.28	19.55329
HNS	1	4.086614	4.088	−6.51	78.2969
NQ	0	4.626304	3.071	50.4	−98.8132
Explosive D	1	NR	NR	NR	−393.578
TATB	1	3.059524	3.062	19.4	−154.291
DATB	1	3.339637	4.1	14.06	−98.8132
NTO	0	NR	NR	NR	−59.8741

Calculation of Heat of detonation is investigated further by Keshavarz[125]. Rather than relying on empirical simplified decomposition products, it is proposed that decomposition of explosives may occur along four different pathways and this depends on oxygen balance of the compounds or formulations. In the paper, only molecular structure of the explosive is used for calculation of chemical energy release. The empirical relation developed is given below.

Heat of detonation, kJ/g = $3.198 + 1.223\, r_{O/C} - 0.625\, r_{H/O} + 1.193\, P_{cyc\ nitramine} - 1.408\, C_{polar}$, where $r_{O/C}$ is ratio of oxygen to carbon, $r_{H/O}$ is ratio of hydrogen to oxygen, $P_{cyc\ nitramine}$ is correcting figure for cyclic nitramine (1 for RDX), C_{polar} is polar or functional group contribution (0.8 for aromatic energetic compounds like –COOH, $-O^-$, two –OH, one –OH and one $-NH_2$, $3-NH_2$, 1.25 for nitramines

[125] M.H. Keshavarz, "A Simple Way to Predict Heats of Detonation of Energetic Compounds only from Their Molecular Structure. *Propellants Explosives Pyrotechnics*, Vol. 37, No. 1, Jan 2012, pp. 93–99.

with polar group $-NH-NO_2$, 2.5 for presence of nitrate salt of amino group ($-NH_2.HNO_3$).

For RDX, $r_{O/C} = 2$, $r_{H/O} = 1$, $P_{cyc\ nitramine} = 1$, $C_{polar} = 0$. Hence, heat of detonation = 3.198 + 1.223 × 2 – 0.625 × 1 + 1.193 × 1 – 1.408 × 0 kJ/g = 6.212 kJ/g = 1.4836 kcal/g. The value matches with results in Table 2 for type I method of calculation for RDX (1481.05 cal/g).

The process is repeated for TNT, where $r_{O/C} = 6/7$, $r_{H/O} = 5/6$, $P_{cyc\ nitramine} = 0$, $C_{polar} = 0$. Hence, heat of detonation = 3.198 + 1.223 × 6/7 – 0.625 × 5/6 + 1.193 × 0 – 1.408 × 0 kJ/g = 3.7254 kJ/g = 0.8897 kcal/g. The value matches with results in Table 2 for type II method of calculation for TNT (977.68 cal/g).

The method is although empirical but the consistency in equivalence with earlier proposed methods is missing. Sometimes, it matches with type I calculations and sometimes with type II. In fact for NTO, calculated values of heat of detonation is reported as 2.856 kJ/g, which is equivalent to 0.6821 kcal/g. The value matches with results in Table 2 for type II method of calculation for NTO (741.045 cal/g). the calculations are also presented in the paper for all explosives of Table 1 along with certain advanced explosives like CL-20, FOX-7 etc.

8

TNT Equivalence

Strength of explosive is very relevant for explosive molecules and formulations. Although heat of explosion is one parameter, but output from an explosive is measured in terms of mechanical work that it does on the surrounding. This mechanical output of an explosive responsible for expansion of product gases is a part of total heat of explosion of the explosive. In this connection, Berthlot approximation is an approach. For this product of explosion of the explosive based on scheme derived in previous chapter is used. The change in number of moles of gaseous product and heat of explosion is calculated.

Strength $\propto \Delta n\ \Delta E/MW^2$, where Δn = change number of moles of gas per mole of explosive on chemical reaction, ΔE = heat of explosion of explosive, MW = molecular weight of explosive.

The approach desires use of TNT as reference for measurement of relative explosive strength. For TNT, molecular formula is $C_7H_5N_3O_6$, which has molecular weight of 227.158 (Table 2) and heat of formation of –15 kcal/kg (Table 1). The associated chemical reaction indicates following gaseous products – 3/2 N_2 + 5/2 H_2O + 7/2 CO + 7/2 C. Total number of moles of gaseous molecules formed = 3/2 + 5/2 + 7/2 = 15/2. Since explosive is solid, change in number of gaseous moles is 15/2 (Δn). Heat of formation of CO and H_2O are –26.46 kcal/mol and – 57.79 kcal/mol respectively. N_2 has zero heat of formation. Change in energy, ΔE = [–15 – (5/2 × – 57.79 + 7/2 × – 26.46)] kcal/mol = 222.085 kcal/mol. So, strength of TNT $\propto$ 15/2 × 222.085/227.158^2 = 0.0322793. The relative strength is expressed in term of this numerical value. Strength of explosive divided by strength TNT multiplied by 100 gives relative strength in percentage (%).

Relative strength = 100 × [Δn DE/MW^2]explosive/[Δn ΔE / MW^2]$_{TNT}$ = 3097.9584 × [Δn ΔE / MW^2]$_{explosive}$

For RDX, molecular formula is $C_3H_6N_6O_6$, which has molecular weight of 222.174 (Table 2) and heat of formation of 14.71 kcal/kg (Table 1). The associated chemical reaction indicates following gaseous products – $3N_2$ + $3H_2O$ + 3CO. Total number of moles formed = 3 + 3 + 3 = 9. Since explosive is solid, change in number of gaseous moles is 9 (= Δn). Heat of formation of CO and H_2O are –26.46 kcal/mol and –57.79 kcal/mol respectively. N_2 has zero heat of formation. Change in energy, ΔE = [14.71 – (3x – 57.79 + 3x – 26.46)] kcal/mol = 267.46 kcal/mol.

So, relative strength of RDX = 3097.9584 × [Δn ΔE/ MW^2]$_{RDX}$ = 3097.9584 × 9 × 267.46/222.1742 = 151.07%

Similar calculation is repeated for HMX. For HMX, molecular formula is $C_4H_8N_8O_8$, which has molecular weight of 296.232 (Table 2) and heat of formation of 17.93 kcal/kg (Table 1). The associated chemical reaction indicates following gaseous products – $4N_2$ + $4H_2O$ + 4CO. Total number of moles formed = 4 + 4 + 4 = 12. Since explosive is solid, change in number of gaseous moles is 12 (= Δn). Heat of formation of CO and H_2O are –26.46 kcal/mol and – 57.79 kcal/mol respectively. N_2 has zero heat of formation. Change in energy, ΔE = [17.93 – (4 x – 57.79 + 4 x – 26.46)] kcal/mol = 354.93 kcal/mol.

So, relative strength of HMX = 3097.9584 × [Δn ΔE/MW^2]$_{RDX}$ = 3097.9584 × 12 × 354.93/296.232^2 = 150.36%

This calculation depends on probable product of explosion, which affects change of number of moles as well as total heat. There are many schemes for predicting product of explosion for calculation purposes and each scheme gives different results. This relative strength has limited use.

As per another scheme of decomposition of molecules, TNT and RDX decompose as given below.

$$C_7H_5N_3O_6 \rightarrow 1.5\ N_2 + 6\ CO + 2.5\ H_2 + 1\ C$$

For TNT $\Delta n = 1.5 + 6 + 2.5 = 10$. Heat of formation of CO is – 26.46 kcal/mol. H_2 and N_2 have zero heat of formation. Change in energy, $\Delta E = [-15 - 6x - 26.46]$ kcal/mol = 143.76 kcal/mol. So, strength of TNT $\propto 10 \times 143.76/227.1582 = 0.02786$.

Relative strength = $100 \times [\Delta n\ \Delta E/MW^2]_{explosive}/[\Delta n\ \Delta E/MW^2]TNT = 3589.37 \times [\Delta n\ \Delta E/MW^2]_{explosive}$

The calculation is repeated for RDX and HMX also. Since there is no change in product of explosion, explosive parameters for explosives will not change from previous calculations.

So, relative strength of RDX = $3589.37 \times [\Delta n\ \Delta E/MW^2]_{RDX} = 3589.37 \times 9 \times 267.46 / 222.174^2 = 175.04\ \%$

So, relative strength of HMX = $3589.37 \times [\Delta n\ \Delta E/MW^2]_{RDX} = 3589.37 \times 12 \times 354.93 / 296.232^2 = 174.21\ \%$

The relative explosive strength gives some insight into explosive behaviour assuming TNT as reference molecule. Output of an explosive is denoted in explosive industry as 'TNT Equivalence". It is a measure of power of explosives and explosives formulations with respect to TNT. It has several regulatory and research oriented utilities. The importance of TNT equivalence is listed below.

- It is used for limit imposed upon size and location of explosive storage facilities
- It is used for citing and construction of explosive manufacturing facilities
- It is also used for packaging and transportation of explosives
- It is used for scaling calculations for air-blast, cratering, structural damage, physiological effects

Selection of TNT as reference for such calculations is based on several facts.

- TNT is one of the main ingredients in all-explosive formulations.

- TNT shows repeatable performance.
- Pure form of TNT can be easily made available.
- TNT can be melted and resolidifed without loss of performance.
- Making TNT charge is easier.
- Data on performance of TNT charges are available in plenty.

This performance parameter term is based on practical performance in terms of explosive effects like creation of blast waves, craters, structural response etc. However, TNT equivalence is measured by different tests and consistency of these experimental observations for different classes of explosives is sometimes evasive. Various test methods include Plate dent test, Ballistic mortar test, Trauzel lead block test, Sand crush test, Air blast test[126]. The output parameters measured as per different tests are different and some of the properties used for calculating TNT equivalence of explosives are given below.

- Chapman-Jouget (C-J) pressure of explosive
- PV-work in C-J zone
- Area under blast pressure-time curves

TNT equivalent is sometimes referred as property of a given explosive. However, there are several controlling factors, which give variable TNT-equivalent for the same explosive. The value depends on nature of explosive and its visible effect on detonation. An explosive can have TNT-equivalent higher than or lower than 1. Typically ammonium nitrate based explosives have TNT equivalent less than 1. Explosives on detonation shows different effects like blast pressure, impulse, brisance, energy release, sound, flash etc. Any one of them can be taken for comparison and accordingly TNT-equivalent can be assigned. It is not effect alone, the distance from explosion site is also important for TNT-equivalent and different distances exhibits different TNT equivalent for same substance for same detonation effect. The

[126] Paul W. Cooper, "Comments on TNT Equivalence", 20th International Pyrotechnics Seminar, Colorado Springs, Colorado, July 24–29, 1994.

fourth parameter, responsible for variation is shape of explosive charge. A spherical charge emits symmetrical waves or effects but cylindrical or other shapes give asymmetrical detonation waves. The location of initiation also plays a role in assessment of TNTequivalent. So overall TNT-equivalent cannot be treated as explosive property and is affected by the following five factors[127]:

- Nature of explosive
- Explosive effect considered for comparison
- Distance of measurement from site of explosion
- Shape of the explosive charge
- Location of initiation

A detail of various tests are given in brief below, which are basically different explosive effects of detonating explosives, simulated for comparison with same effect of TNT and calculation of TNT-equivalent.

***Sand crush Test*:** In this test, given explosive, initiated by lead azide or mercury fulminate or tetryl initiator is allowed to crush sand and sand crush value is obtained by subtracting weight of sand crushed with initiator alone from weight of sand crushed by explosive initiated by initiator. In this test, generally 0.4 grams explosive sample is taken and it is pressed at around 20.6 MPa. The sand is fine Ottawa sand finer than 30 mesh. The quantity of initiator is variable and that needs subtraction of weight of sand crushed by initiator alone. Sand crush value is net weight of the sand crushed.

Crushing or shattering of sand depends on shock pressure incident on it. Pressing load of explosive is specified in this test, but different densities are achieved with different explosives for the same crushing load. Since different densities for same explosive give different shock pressure, comparison of different explosives maintained at different densities is not a very logical test. In

[127] Bart Simoens, Michel H. Lefebvre, "Influence of Different Parameters on the TNT-Equivalent of an Explosion". *Central European Journal of Energetic Materials*, Vol. 8, No. 1, 2011, pp. 53–67.

addition, there is no mechanism to transfer all the energy of explosive to sand. Some part of expositive energy is always lost in shock heating, expansion, fracturing and acceleration of container components. These forms of losses are linearly proportional to shock pressure and are not constant for each explosive. Sand crush value (gram of sand) for RDX and HMX is found to be around 60, despite minor difference between their heats of explosion (part of chemical energy available for detonation). The TNT equivalence value is around 136%. Higher heat of explosion generally gives higher TNT equivalence by sand crush value, but trend is not very universal. The values do not represent explosive power or any energy characteristics, where rate effects are accounted. If only shattering ability of explosives is accounted then C-J pressure is a better measure, which can be accurately estimated at known density and detonation pressure.

In place of sand, sometimes lead-cylinders (diameter 40 mm × height 60 mm) are also used. It is placed over an unyielding base, which acts as shock pressure attenuator. On detonation of explosive charge, the lead cylinder will be compressed and a mushroom head will be generated at the top. The reduction in height of lead cylinder can be taken a direct measure of brisance of the explosive, which can be correlated with that for TNT. Generally, TNT with 1.2 g/cc density reduces height of lead cylinder by 18.8 mm through a 10 mm thick steel disc (placed between explosive and lead cylinder). Hess proposed this test in 1876.

Another way to measure brisance or shattering power of explosive is proposed by Kast. In this case compression of copper crusher is used as control parameter. Explosive charge is prepared in a zinc tube (internal diameter 20 mm × length 80 mm × wall thickness 0.3 mm). The copper crusher is made of electrolytic copper and is of 7 mm diameter and 10.5 mm in height. When explosive is initiated, it presses copper crusher, which is measured as reduction in height of copper as units of crushing.

Trauzel lead block test: This is a test to measure capacity of explosive to expand a cavity in lead block. Generally 10 grams of explosive is exploded in a cavity of 25 mm diameter and 125 mm deep in a lead block of 200 mm diameter and 200 mm height. Cavity should be drilled centrally at the top surface. The quantity of explosive and initiator are adjusted so that total expansion is restricted between 250 cc and 300 cc. The test values are expressed in percent of the expansion of an equivalent weight of TNT. The initiators like lead azide, lead styphnate, mercury fulminate etc are found to show around 50% TNT equivalence. This means their strength is half that of TNT. Nitroguanidine has lower heat of explosion than TNT but it has 100% TNT equivalence. It is observed that using this test RDX is more powerful than HMX. The value of RDX is around 186%. Nitroglycerin and PETN exhibit higher TNT equivalence than HMX.

The lead block used in this test is made from soft lead, cast at a temperature of around 400°C. Sometimes, rather than exploding 10 g explosive charge, a constant volume of 10 cm^3 is used as standard. The expansion of cavity is measured by pouring water in the expanded cavity and subsequent measurement in a graduated measuring cylinder. The explosion in the cavity raises temperature of cavity. If temperature rise is within 20°C, no correction is applied else corrections are applied. For more rise, negative correction is applied. Before testing any new explosive sample, calibration of lead block is mandatory. The calibration is generally done by pure picric acid at density of 1 g/cc and charge volume of 10 cm^3. The net expansion of lead cavity must be within 287 cc to 300 cc.

Similar to sand crush test, Trauzel lead block test also reply on C-J pressure. Unfortunately in this test, neither density is controlled, measured, mentioned or reported. So this test also suffers from unknown incident shock pressure. The expansion of cavity in lead is a hydrodynamic process and springing ability comes into picture for such tests. This test can also be duplicated by estimation of C-J pressure.

***Ballistic mortar test*:** This test resembles charpy or izod impact test and is basically lifting a pendulum type weight by swinging to higher levels. In this case performance of test explosive is compared with 10 grams of TNT. Ballistic mortar is a long supporting rod at the end of which a heavy short-nose mortar is supported in hanging condition. The mortar has a 150 mm diameter and 300 mm long chamber in which 175 mm is occupied by a projectile. Sample is positioned in the remaining part of the chamber and when detonated, projectile is thrown in a sand bank and mortar swings. The maximum swing angle is measured and height is compared with firing of 10 grams TNT charge under similar condition to get TNT equivalence. RDX and HMX exhibit same TNT equivalence by this method. The value of RDX is around 140%. PETN and NG exhibit lower TNT equivalence than HMX (or RDX) and this trend is reverse of that obtained by Trauzel lead block test. When TNT equivalence by this method is compared with C-J detonation pressure, no trend emerges.

Energy from detonation of explosive is consumed in two actions – one is by movement of projectile and other by movement of the ballistic pendulum. First part of the energy can be accounted by taking kinetic energy of the projectile, while the second part is calculated by potential energy of the pendulum. The ballistic mortar gets initial velocity from projectile and the gain in its potential energy can be equated to kinetic energy imparted by the projectile. From this velocity of the projectile can be calculated. The explosive strength is proportional to (1-cosA), where A is the angle by which ballistic mortar rises due to explosion.

When explosion occurs very near to the projectile, there are chances of shock heating and plastic deformation. These are basically loss of energy of explosive in unintended purpose. These losses of energy are dependent on C-J pressure and higher pressure leads to higher losses. Additionally, some part of energy is wasted during expansion of hot gases to propel the projectile before mortar starts swinging upward. Once projectile is out of mortar, sudden expansion causes no further application of force on the mortar. In some explosives, reaction continues to release energy rather slowly. Aluminized explosives come in this category and their later release of energy after projectile exit is not accounted. In this test also density effects are not considered and comparison of performance seems inadequate. The results obtained by this method are specific to test instruments and are not representing properties of explosives in other configuration or environment. The calculation based on conservation of linear momentum alone is probably not sufficient to measure explosive strength and role of angular momentum cannot be ruled out.

Plate dent test: In this test, capacity of explosive charge to form dent in a steel plate of given thickness is assessed. The explosive charge is unconfined having 40 mm (13/8 inch) diameter and 125 mm (5 inch) length. It is fired on a 45 mm (7/4 inch) thick 125 mm (5 inch) square plate of cold-rolled steel plate. Some backing plates are also provided. The output of the result comes in the form of depth of dent in absolute terms. Sometimes this is compared with dent created by TNT charge at a density of 1.61 g/cc. The diameter and lengths of the charges are variable. Since dent produced scale well with charge dimension, charge dimensions do not affect the relative measure of output. However length to diameter (L/D) ratio of the charge affects output. For L/D ratio lower than 2, depth of plate dent increases with increase in L/D ratio. If L/D is more than 2, depth of dent attains maximum value. For L/D more than 2, scaled dent varies linearly with C-J pressure and plate dent test can be used to give relative values of C-J pressure. The test can be used to determine elastic/plastic behaviour of materials like yield strength of the materials also.

Since steel plate is used as witness plate, proper calibration of this plate is mandatory. For calibration, depth of dent for different densities of TNT is considered and equation of the form depth of dent, h (mm) = A $\times$ ρ (g/cc) + B is developed. Generally, higher density means higher value of dent depth and 'B' is a negative coefficient. For TNT of density 1.633 g/cc, depth of plate dent is observed as 6.706 mm[128]. The relative brisance is calculated as depth of dent for the trial explosive divided by same for TNT, expressed in percentage. For RDX at density of 1.767 g/cc, the depth of dent is of the order of 10.516 mm, giving TNT equivalent as (10.516 $\times$ 100/6.706 =) 156.8. Similarly for HMX at density of 1.730 g/cc, it is 10.084, giving TNT equivalent of 150.4. For Tetryl, at density of 1.681 g/cc, the dent depth is 8.103 mm, giving TNT equivalent of 120.8. In fact this test is based on fact that brisance is

[128] Smith L.C., "On Brisance and Plate-Denting Test for the Estimation of Detonation Pressure". *Explosivstoffe*, Vol. 5, 1967, pp. 106–110.

dependent on detonation pressure of the explosive. After exhaustive trial and calibration following results are developed for dent depth and detonation pressure.

Detonation pressure, P (kbar) = 33.374 × h (mm).

Cylinder expansion test: Actual strength of an explosive or its ability to do useful work is derived, if measurements are made on the basis of actual expansion work of a metallic obstruction. For this cylinder expansion test is very useful and it simulates detonation in an actual warhead, where expanding product of explosion destroys cylindrical casing. For each explosive Gurney energy ($\sqrt{2E}$) is a measure of explosive effects on the surrounding. It is kinetic energy of expanding detonation products and kinetic energy of the displaced material (wall of the cylinder).

Velocity of cylinder = $\sqrt{2E} / \sqrt{(M/C + ½)}$, where C = explosive quantity, M = mass of the metal cylinder, $\sqrt{2E}$ = Gurney constant of the explosive.

If total energy of the explosive (product of mass of explosive and heat of detonation) is consumed for expansion of cylinder wall, maximum wall velocity will be realized. The energy of explosive can be equated to kinetic energy of the metal wall. The heat of detonation (Q) can be give detonation velocity ($D = \sqrt{[2 (\gamma^2 - 1) Q]}$. For $\gamma = 3$, $D = 4 \sqrt{Q}$. If some part of total explosive energy is spent in imparting kinetic energy and heat to the detonation products, maximum wall velocity will reduce slightly.

Maximum velocity of the cylinder = $D/2 \times \sqrt{[C/(C + 2M)]}$, where D = detonation velocity of the explosive

The cylinder expansion is generally measured by optical means at a given section with respect to time. Generally a copper cylinder (diameter 25.4 mm × length 300 mm × wall thickness 2.6 mm), filled with explosive is considered. Explosive is initiated at one end and expansion in diameter of the cylinder is monitored at a given cross-section. To attain steady flow of cylinder wall, increase in radial expansion by 5 mm is taken as start point and point of maximum velocity is taken equivalent to radial expansion of 19 mm.

The output of this test may be in the form of cylinder expansion curve, wall velocity – time curve, acceleration-time curve or isentropic expansion curve of the explosive.

***Air blast test*:** In this test, blast pressure and impulse at different distances from encased explosive, exploded in open atmosphere is measured and compared. The explosive is encased in steel container and is generally cylindrical in shape. The explosive charge is kept at certain height from the ground for two reasons – (*i*) loss of energy due to cratering (*ii*) minimization of error due to ground reflection. The weight of casing and explosive is taken near one. In air-blast experiment, blast pressure or impulse (integration of pressure versus time profile) is considered for calculation of TNT equivalence.

The blast can be explained in terms of severe mechanical loading on the structures in the vicinity of explosion cites. An explosive blast resulting from an explosion has a sudden forceful blow of air from the instantaneous pressure jump. This has a short life. This is followed by the crushing effect of blast overpressure (pressure above atmospheric) and a blast wind of very-high velocity. These blast wave effects decrease quasi-exponentially with time, until the pressure reaches atmospheric (zero overpressure), further to less than atmospheric pressures. The duration of negative overpressure (phase with less than atmospheric pressure) is longer. This is followed by a reversed blast wind. Blast waves are characterized by the initial shock intensity (peak overpressure, mach number, or particle velocity), duration of the blast wave (positive overpressure phase), and impulse (force-time product) per unit area for the pressure forces in the blast. The arrival of the shock front from the center of the explosion to the location of concern is also an important parameter.

However, it is difficult to correlate air blast test results with other mechanisms for calculation of TNT equivalence. When explosive charge is exploded, energy is transferred to metal casing in three modes – (*i*) shock heating, (*ii*) strain or fracture and (*iii*) kinetic energy of fragments. Detonation produces shock of around 300 kbar in steel and temperature of steel may go to 300°C. The energy lost is around 35 cal/gram of steel. The strain or fracture consumes energy of the order of 25 to 150 cal/gram. If fragment is launched at a velocity of 1.9 km/s, the kinetic energy will be around (= $\frac{1}{2}$ 10^{-3} × 1900^2 J/g) 430 cal/g. This energy is unavailable to the blast wave. The PV-work energy available from TNT is about 1160 cal/g, out of which around 500 cal/g is consumed by steel. So, air-blast experiment represents only remaining 660 cal/g of energy, which is around half the complete available energy. As C-J pressure increases, energy lost in shock heating also increases. Energy lost to strain depends on strain rate, which is again a function of C-J pressure. Energy consumed in fracturing is a function of newly created surface area. If larger areas are created, more energy will be consumed. High C-J pressure produces smaller fragments and hence larger area. So, energy loss by this mode is also on higher side for higher C-J pressure of explosive. As C-J pressure increases, energy lost to impart kinetic energy also increases. So, higher C-J pressure leads to higher losses of energy. The TNT equivalence by this method depends on residual energy of explosive after certain loss of energy to steel casing. If an explosive has more energy than TNT losses are more and residual energy will not be as high as it should be and so is nature of TNT equivalence.

Another anomaly in air-blast based prediction of TNT equivalence is that it underestimates TNT equivalence of aluminized explosives. Actually for ideal explosives, chemical reaction should be completed in C-J zone or reaction zone, but for non-ideal explosives, this continues behind C-J zone. For such explosives, energy is released at rate slower than ideal explosives. In aluminized explosives, chemical energy release takes place much behind C-J zone, which is not included in prediction approach. So, predictions for non-aluminized fail to estimate and include contributions of slow burning of explosive. However, if heat explosion is considered, the TNT equivalence is highly overestimated for aluminized explosives. This overestimate is because, all the present aluminium is not consumed in detonation and only a small portion of aluminium burns at high-pressure to contribute to the blast waves. So calculation for aluminized explosives by this scheme is also not reliable. The complete detonation phenomena till date are not modeled completely by any of the advanced hydrocodes.

The blast pressure prediction mechanism is also very difficult. The peak pressure varies with distance from the explosion site and so is rate of pressure decay. If peak pressure is high rate of decay will be high and positive pressure phase will be lower. As peak pressure reduces, rate of pressure decay reduces and duration of positive pressure phase increases. This effect says that when an explosive gives higher peak pressure than TNT, its decay rate will be faster and after certain distance observed pressure of blast wave from explosive may be lower than that from TNT. This indirectly says that ratio of blast pressure will vary with distance and unique value cannot be assigned to TNT equivalence by this method. This can be said to be an offshoot of non-identical blast pressure waves by different explosives.

The air blast data can be utilized for assessment of sympathetic detonation of explosive stores also. It is found that sympathetic detonation is dependent on detonation properties of donor, the delivered forms of explosion energy and sensitivity of acceptor. Blast overpressure for TNT is practically determined by conducting test of cylindrical charges (mass – W kg) of TNT and measuring blast overpressure at several distances (R in m). A third order polynomial fit is given below[129].

$P\ (\text{MPa}) = 158.3 / Z + 15.3 / Z^2 - 33.7 / Z^3$, where $Z = R/W^{1/3}$ is scale distance.

Using experimental results, R_{100} (maximum distance from donor explosive at which sympathetic detonation occurs) and R_0 (minimum distance from the donor at which acceptor cannot be detonated) for charge density of 0.85 g/cc is calculated in the report and is reproduced below.

$$R_{100} = 0.295 \times W^{0.39}$$
$$R_0 = 0.420 \times W^{0.29}$$

[129] Xiang Xuzhang, Guo Zixi, Zhao Zhuanghua, Yang Keyi, "Studies on Sympathetic Detonation of Bulk TNT in Large Quantity", Proceedings of International Symposium on Pyrotechnics and Explosives, Beijing, China, Oct 12-15, 1987, pp. 713–718.

However, engineering application require a simpler relation for the same, as given below.

$$R_{100} = 0.5 \times W^{1/3}$$
$$R_0 = 0.7 \times W^{1/3}$$

TNT equivalence is a method to define explosive energy with TNT as reference. For plastic deformation, cratering or shattering effects, C-J pressure gives better estimate than TNT equivalence. This can be easily and accurately measured and estimated for known density and velocity of detonation. Plate dent test to some extent replicates C-J pressure. Prediction of blast and its structural and physiological effects, TNT equivalence is definitely a better tool.

***TNT equivalent of Propellants*:** TNT equivalent is used for safety certification of advanced propellants also. The mechanism of development of detonation in granular and crystalline explosive due to thermal initiation is explained through pressure development in confined space. The thermal initiation forms gases, which fails to escape from the crystals and pressure gradient develops. This has synergistic effect of increase of burning rate and pressure in accelerated manner. This condition forms shock waves which are reinforced by the energy released by the burning explosive and they eventually reach an intensity where the energy of the reaction is used for propagation of the shock wave and a stable detonation front is produced. A critical mass exists for each material above which this deflagration can pass over into detonation under proper conditions. Below the critical mass, the burning will increase and then decrease as the material is consumed.

There is definite evidence that the process of transition from deflagration to detonation (DDT) for explosives and propellants is a continuous reaction consisting of

- ignition,
- pressure rise in confinement,
- pre-detonation reaction at accelerated burning rate
- fast rate of change of pressure

- establishment of temperature gradient
- development of an accelerating shock front
- detonation of sufficient mass of material

It is believed that any material which can be detonated should exhibit this pre-detonation reaction. In the case of very sensitive primary explosives the level of controlling parameters required to start detonation is so low that they cannot be measured by present techniques. For "non-detonable" propellants the pressures required for the pre-detonation reaction to occur are so high that for all practical purposes, they cannot be attained. It is considered practical that this technique can be used for the classification of the detonation hazard for a particular motor configuration if the pre-transition pressure and slope of the burning rate pressure curve of the propellant used is known. Thus, for example, if a defect or void should exist in a propellant, which might conceivably ignite on firing, by considering such an ignition as an interior ballistic system the pressure and rate of pressure rise can be calculated to determine if pre-detonation conditions could develop before tensile failure of the grain occurred. If such reaction can occur then the accelerated pressure rise could develop the shock front necessary for transition to detonation. Definitely exploding propellant is a challenge and both mass and critical diameter criteria have to be met for a possible detonation of propellant.

TNT equivalence of ball powder composition was determined[130]. Scaled positive impulse TNT equivalency values for the ball powder in shipping container were greater than 100% at the near field values, *i.e.* less than 7.14 m/kg$^{1/3}$. However, the values were less than 100% at scaled distances greater than 7.14 m/kg$^{1/3}$. Reverse trend is observed, when TNT equivalency was determined in dryer bed configurations. Prior to this, similar analysis was conducted for M10 propellants[131]. It was observed that TNT equivalency

[130] F. L. Mcintyre, TNT Equivalency of Ball Powder WC844, US Army Armament Research And Development Command, Contractor Report ARLCD-CR-80024, September 1980.

[131] F. L. Mcintyre, Paul Price, TNT Equivalency of M10 Propellants, US Army Armament Research And Development Command, Contractor Report ARLCD-CR-78008, March 1978.

values were greater than 100% at the near field values, *i.e.* less than 3 m/kg$^{1/3}$ and more than 100% at far field (>3 m/kg$^{1/3}$). The trend was similar to that for the ball powder. Later on, TNT equivalency of nitrocellulose on the basis of pressure was also determined. Although observations were similar to other previous calculation, some anomalies were also observed[132]. In terms of impulse, TNT equivalency was always less than 100%. TNT equivalency values of nitrocellulose under simulated conditions found in the Thermal Dehydration Unit were generally less than 100% for pressure and scaled impulse. The exceptions were at the near field value of 1.19 m/kg$^{1/3}$ where the pressure equivalency was 130%, and the far field value of 7.14 m/kg$^{1/3}$ where the impulse equivalency was 105%. In one of the latest report, many gun propellant formulations were analyzed for blast overpressure and TNT equivalent were determined. Tests were conducted for M1, M6, M30A1 gun propellants of US army. All the compositions were initiated by C4 pellets in their respective containers. M1 and M6 showed lower than 100% TNT equivalency for pressure and impulse values of air blast, while M30A1 showed a higher than 100% TNT equivalence[133].

There is as yet no adequate theory to predict the critical diameter of an explosive system. In an effort to determine the detonability of Polaris, using composite propellants, a series of tests, code named Beauregard, was performed during the summer and autumn of 1958 at the Naval Ordnance Test Station. It was established that the critical diameter for detonation of the solid non-defective composite propellant was above 20 inches. The implication of these tests is that motors of web thickness of the order of 20 inches or less are not detonable. On the other hand, considerable blast effect, attributable to the propellant, was recorded, which indicates that, in the presence of a severe shock, a significant quantity of the propellant can react rapidly enough

[131] F. L. Mcintyre, P. Price, TNT Equivalency of Bulk Nitrocellulose, U.S. Army Armament Research and Development Command, Contractor Report ARLCD-CR-81007, March 1981.

[132] Michael M. Swisdak, Jr., TNT Equivalence of Selected U.S. Army Gun Propellants, Naval Surface Warfare Center, US, August 1994.

to contribute to the peak overpressure. For a given shock donor and with truly non-detonable acceptors (like propellants), it has been established that the measured TNT equivalent decreases with increasing length of the propellant sample. Thus, though this test does relate to establishment of shipping criteria for these small samples, the TNT equivalents calculated from peak pressure measurements cannot be extrapolated to larger systems, especially those of complex geometry.

Since propellant cannot be detonated, explosives were added in propellant formulations for creation of blast in so called adulterated propellants. To assure detonation, the initial test specimens were detonated with an excessively large quantity of TNT. By gradual reduction of RDX enrichment and increasing test specimen diameter, it was possible to identify a critical diameter of approximately 64 inches for a composite propellant right solid circular cylinder. Additional tests were performed to determine the minimum shock pressure required to initiate detonation as a function of propellant diameter. Results obtained from RDX adulterated propellants and extrapolated to unadulterated propellants indicated that Class 2 solid motors near their critical diameter would mass detonate when exposed to an overpressure of 25 kilobars or greater.

Although there exists considerable difference in terminal yield equivalency data, it can be seen from the Minuteman 3rd stage data that the TNT equivalency can be considerably higher at close range approaching a lesser value at larger distances. Terminal yield TNT equivalency for composite modified double base propellant motors was 130-140%. Composite propellants near their critical diameters were capable of explosive yields of 156% to 168% based on combined overpressure and impulse data. These propellants can show higher yields (197%) based on peak side-on overpressure only. Actual tests of smaller rocket motors indicate that these yields are rarely achieved and that composites achieve yields in the order of 85% TNT equivalency while double base propellants can achieve yields up to 140% TNT equivalency. Combined tests of double base and composite propellants in which the former was used as the donor achieved yields in the order of 125% TNT equivalency[134]. In another report depicting project Sophy for composite propellants, the critical diameter of a typical composite propellant was determined to be between 60 and 72 in. The minimum shock pressure to initiate composite propellant sample at critical diameter was estimated to be 25 to 30 kilobars. The average TNT equivalence of the peak pressure data is 197%. The average TNT equivalence of the propellant based on impulse data is 138%, for those samples that detonated[135].

In another series of tests, propellant for Titan III was tested. The propellant was based on HTPB, AP (69%) and Al (19%) and solid loading was of the order of 88%. The burning rate of the propellant was around 8 mm/s at 76.9 MPa. The TNT equivalent of propellant was mentioned in the report as 1.3, while that for C4 was mentioned as 1.19. An explosive blast effect similar to combination of both was observed when they were fired in conjunction. 5 kg C4 (= 5.95 kg TNT) and 48 kg propellant (=62.4 kg) showed same effect as shown by 68 kg (~ 5.95 + 62.4) of TNT. High overpressure yield was obtained despite observation of some burning propellant pieces being thrown clear from ground zero. Since no detonation of propellant had occurred, the test demonstrated that slow burning, aluminized, HTPB propellant could give high atmospheric overpressure shocks without being involved in detonation behavior[136].

For the determination of TNT equivalent, an ab-initio theoretical modeling based on minimum free energy method was also explored[137]. This does not require any practical field testing

---

[134] D. F. Pennington, T. Man, B. Persons, Rocket Propulsion Hazard Summary: Safety Classification, Handling Experience and Application to Space Shuttle Payload, NASA report - NASA CR-145185, May 1977.

[135] Elwell, Robert B., Irwin, Owen R., Vail Jr., AT Richard W., Project SOPHY - Solid Propellant Hazards Program, Aerojet General Corporation, AFRL-TR-67-211 – Vol. II, Aug 1967.

[136] Claude Merrill, Albert Nichols, Edward Lee, Explosive Testing of Class 1.3 Rocket Booster Propellant, August 1994.

[137] Ning He, Cong Xiang, Bin Qin, Qi Zhang, Calculation of TNT Equivalence of Composite Propellant and Visualized Software Development, *High Performance Computing and Applications*, Vol. LNCS 5938, 2010, pp. 161–168.

of propellant in detonation mode. The chemical balance is achieved when total free energy of the species attains a minimum. The propellant containing 70% Ammonium perchlorate (AP), 18% aluminium (Al) and 12% HTPB, can give explosion heat of 3.616 MJ/kg and has an TNT equivalence of 0.798. Another propellant formulation with 60% AP, 28% Al and 12% HTPB gave explosion heat of 4.05 MJ/kg and TNT equivalence of 0.89. This indicates that replacing AP by Al increases TNT equivalence of the composition. RDX based formulations were also analyzed by free energy minimization method. A composition containing 70% AP, 15% RDX, 5% Al and 10% HTPB gave explosion heat of 4.36 MJ/kg and TNT equivalence of 0.95. Another composition with high amount of RDX, even with reduced solid loading gave higher TNT equivalence. The composition contained 47.4% AP, 32.8% RDX, 6% Al and 13.8% HTPB. It gave explosion heat of 6.74 MJ/kg and TNT equivalence of 1.48.

For Space shuttle, elaborate program was undertaken to assess the hazards involved with propellants. The major concern was situation when liquid fuel (LOX + LOH) tank burst at launch pad. The space shuttle in fully loaded condition on the pad contains approximately 4 million lb of LOH and LOX propellants. Of these, 2.9 million Ib of LOX and 0.5 million Ib of LOH are stored in two tanks aboard the booster. The orbiter has two LOX tanks with a capacity of 0.25 million Ib each and a single LOH tank with 0.1 million Ib capacity. The current criteria specify that the TNT equivalence for LOX-LOH propellant combinations is 60%. Based on this equivalency, the facilities for the space shuttle would have to be based on a total yield equivalent to 2.4 million Ib of TNT.

9

Measurements of Explosive Parameters

Measurement of explosive parameters like detonation velocity, detonation pressure, sensitivity parameters, TNT equivalent, plate velocity etc desires specialized set-up and the calculations are inherently embedded in the experimental set-up. Although many measurement techniques are discussed in respective sections, some of the popular methods for measurement of performance parameter of explosives are illustrated along with computation background.

Dautriche method is the oldest and less resource consuming method for the determination of detonation velocity. In this case, known length of a detonating cord, detonation velocity (D) through which is known is taken and connected to the explosive charge with unknown detonation velocity. Both ends of the detonating cords are mounted at some known distance on the explosive (say L ~ 300 mm). The free loop of the detonating cord is pasted or mounted on a steel or brass witness plate. When explosive charge with unknown detonation velocity is initiated from one end, detonation front establishes (keep distance of around 100 mm between initiator and distance of nearest joint with detonating cord). As soon as detonation reaches nearest end of thc detonating cord, detonation continues in explosive as well as detonating cord. When detonating front in the explosive reaches other end of the detonating cord, detonation moves from other end of the detonating cord also. The detonation fronts, emanating from both the ends of the detonating cord, meet at a point, which is slightly deviated from the center of detonating cord-loop. The meeting point is observed as a mark on the witness plate and its deviation from center point is measured (say t).

Detonation velocity of the explosive = DL/2t

Sometimes, center point is not marked and instead, length of detonating cord on either side of the mark on the witness plate is measured as L_1 and L_2. In this case, detonation velocity is given by $DL/(L_1 - L_2)$.

Since detonation of explosive is associated with generation of light and ionization of gases, these properties are also utilized for measurement of detonation velocity for various explosive charges. However, the equipments become slightly complicated. Nevertheless, ionization probes were used for the measurement of detonation velocity of cast PBX charges based on HMX, Ammonium perchlorate and Aluminium. It was observed that Although density improved, but detonation velocity reduced, for part replacement of HMX by inert AP and Al in PBX[138].

To measure light emission during detonation, high-speed camera becomes an essential requirement. Rotating drum camera is mounted in front of detonating explosive with certain protection against damage from explosion. The film of this camera is mounted on a rotating drum, which rotates about its axis at a predetermined speed. The axes of explosive column and rotating drum are parallel to each other. When detonation wave passes through explosive column mounted in front of objective of the camera, it can be assumed as movement of a light source at high speed in front of the camera. During this movement of detonation front, rotating film also moves by certain distance. If film is not rotating, a vertical line, replicating motion of detonation front is captured on the film. In fact, in a given time, the distance marked on the film is proportional to detonating velocity in explosive. The amplification or otherwise of the distance is given by ratio of distance of explosive and drum from the objective lens. As film is rotating, slant line is created in the film, which can be developed. If film speed is V,

[138] Gordana Antic, Vesna Džingalaševic, Milena Stankovic, Zoran Borkovic, Explosive Characteristics of Cast PBX Based on HMX, Ammonium Perchlorate and Aluminium, *Scientific Technical Review*, Vol. LIV, No. 3-4, 2004, pp. 38–44.

objective to explosive distance is a, objective lens to film distance is b, and angle measured for the slant line from circumferential direction to the drum is θ,

Velocity of detonation of the explosive = $b \times V \tan\theta/a$

Rather than moving film on a rotating drum, streak camera is another device, which amplifies light deviations due to traveling detonating waves. In general, if a mirror is rotated by an angle, the reflected light rotates by twice the angle. In this case, writing speed in km/s can be achieved, which is around one decade higher than that with rotating drum camera. Streak camera was utilized for measurement of velocity of detonation of unconfined explosive charges at small charge diameters[139]. The method was used for assessment of critical charge diameter also. VOD was determined by measurement of the gradient of that portion of the streak record where detonation was stable, taking into account the camera writing speed and the magnification of the image. Those portions at the start of a streak record which showed that explosive was under or overdriven from the EBW detonator were excluded from the measurement. For analysis, an enlarged print was made from each 35 mm streak record. Two methods of analysis were then employed. In the first, a straight line was marked parallel to and just touching the straight line portion of the streak record. A right-angled triangle was then constructed by drawing one line parallel to and one line perpendicular to the fiducial scale. Measurement of these two lines with the Opto-Scale permitted determination of the gradient of the streak line. (This method was found to be more accurate than angular measurement 3f the gradient). As the distance between the fiducial lines and the camera writing speed (corrected for nonlinearity) are known the gradient can be readily converted to VOD. When constructing the triangle a sharp scalpel and a steel straight edge were used to scribe very fine lincs. For measurement, the print was back illuminated by placing it on a light box. The overall measurement accuracy was estimated at ± 30 m/s.

[139] Robert J. Spear and Michael G. Wolfson, Determination of Detonation Parameters of Booster Explosives at Small Charge Diameters MRL Technical Report, MRL-TR-89-45, DSTO Materials Research Library, Australia, 1990.

Pin Oscillographic Technique (POT)[140] is one of the discrete experimental techniques to measure the arrival time of detonation front at pre-marked distances. This is advanced and sophisticated method as compared to Dautriche method of velocity of detonation measurement. It uses high speed oscilloscope and ionization probe. The oscilloscope has frequency in the range of 10 to 100 MHz, which corresponds to a time resolution of 100 to 10 nanoseconds. The probe used in this technique are of shorting type and on arrival of detonation front, which has ionized gases, the probe gets shorted to acknowledge the event. Depending on application the shorting probes may have many varieties – insulated copper wire (0.4 mm dia) bend, twisted and cut at bend can form an open probe-point, which shorts on arrival of detonation front. If media is already conducting, sulfur is coated over the surface of conductor tip. At shock pressure, resistance of sulfur decreases drastically and it becomes conducting. Sometimes, metallic foil probes are used for measurement of jet tip velocity. For measurement plate velocity under explosive loading, single wire probe with common earth is used. Whatever may be the variety of probe, they are mounted at preset distance on the trial sample. The distance between adjacent probes must be around 30 mm for accuracy of measurement and generally in one set 4-5 probes are mounted at equal intervals. From the initiation side, first probe is generally fixed at 100-150 mm or three times diameter of the charge. The measurement of length is by vernier or scale and measurement of time is through oscilloscope.

As a further modification, rather than discrete readings at pre-set distances, continuous determination of detonation velocity is also attempted using special probes. In this case continuous oscilloscope reading of change in electric resistance due to constant current flow and intermittent shorting is recorded and analyzed. Like previous method, the probes are placed parallel to the longitudinal axis of the explosive. Under action of detonation wave, the probe is continuously shortened resulting in change in electrical resistance. This changes voltage, which is recorded on oscilloscope

140 M.N. Adya, Dr. B.M. Bohra, K.U.B. Rao, "Report on Pin Oscillographic Technique", Report No. HEMRL/ED/2003/1, HEMRL, Pune, India, date 27.05.2003.

as a function of time. In this case constant current source is a prime requirement.

Aquarium technique is also debated in explosive literature for determination of detonation parameters. In this explosive charge is placed in an optically transparent media and travel of shock wave is viewed as a function of time. As detonation wave travels through water or plexiglass or other transparent media, the optical transparency is changed, which is recorded by a camera placed perpendicular to propagation of shock wave. The transparent media is illuminated by an impulse high source like Argon flash bomb. Laser is also used for better time resolution and method is almost similar to Aquarium technique. Laser interferometer is used for measurement of detonation parameters and it is based on Doppler Effect. A metal shim is placed between an explosive charge and a transparent material window. On explosion, the metal shim moves and causes Doppler shift in wavelength of the reflected laser beams. Thus position and time is recorded and velocity of shim is estimated.

Electromagnetic particle velocity gauges are also used for measurement of detonation parameters of the explosives. In this case change in electromotive force due to passage of detonation wave is measured. The electromotive force is found to be proportional to the conductor velocity, its size and the magnitude of magnetic field. A thin metal foil gauge placed in the explosive charge is carried away by the detonation products and the motion causes electromotive force. The explosive charge with gauge is placed in a uniform magnetic field. A fast oscilloscope is used to get electromotive force as a function of time behind the detonation wave front. Flash x-ray is a radiographic technique for determination of detonation wave parameter. It is based on time tracing of the position of tin metal foil moving with detonation products.

Although positive pressure phase is considered responsible for damage on a target, damage also depends on the target frequency. If pressure duration of this positive phase is longer than target frequency, peak overpressure is responsible for the damage. Peak overpressure is measured by pressure gauges/transducers with suitable modifications easily. If pressure duration is shorter than frequency of the target, impulse is responsible for the damage. If a 1 kg explosive charge is exploded, the vicinity of explosion site has pressure duration of 1 millisecond only. This is shorter than frequency of most of the targets and for near explosion calculations, impulse becomes the governing criteria for assessment of damage. Direct measurement of impulse is very difficult. However, momentum test method is developed for such measurements. If a momentum gauge is placed at same height as explosive charge, explosion causes push on the gauge. The gauge falls on ground. Fall of 1m takes same time for the gauge and it is independent of the horizontal velocity imparted by the explosion. So, point of fall on ground is directly proportional to initial velocity imparted to the gauge by explosion. The momentum gained by the momentum gauge due to explosion is product of its mass and imparted horizontal velocity. Impulse density is given by momentum divided by projected area of the gauge[141].

Various transducers developed for the measurement of explosives properties are summarized in a paper[142]. Such transducers are mainly used for the measurement of blast parameters. The mechanism of blast is reiterated here again. A forceful blow emerges from the instantaneous pressure jump of the shock front is followed by the crushing effect of blast overpressure (pressure above atmospheric) and an outward blowing wind of super-high velocity. Quasi-exponentially decrease of blast wave with time until the pressure reaches below atmospheric (negative overpressure), along with a reversed blast wind. Blast waves are characterized by peak overpressure, arrival time of blast, duration of the blast wave, and impulse per unit area for the positive overpressure in the blast. The measurement techniques must cover all these parameters.

---

141 M. Held, "Airblast of Different High Explosives in the Damage Distance", 31 DOD Explosive Safety Seminar, San Antonio, TX, USA, 2004.

142 David B. Watts, Michael T. Van Tassel, Transducer Development for Explosive Measurements, Air Force Armament Laboratory, Florida, USA, Report No. AD-A208 615, May 1989.

One common method of measuring the peak overpressure and impulse per unit area of a blast wave is by using free air blast pressure transducers. These transducers are typically piezoelectric (crystal and ceramic) or piezoresistive (strain gage) in nature and do not have the frequency response necessary to accurately measure the transient pressure phenomena. The transducers experience mechanical resonance, which makes it impossible to correctly measure the peak overpressure and, in turn, induces errors in the impulse measurements. Several transducer were developed to achieve sufficient frequency response and resonant frequency for high explosive munitions measurements.

Piezoelectricity is defined as electric polarization produced by mechanical strain in crystals belonging to certain classes, the polarization is proportional to the strain and changes sign with polarization. Piezoelectric pressure transducers employ piezoelectric crystals or ceramics and the performance is greatly restricted in frequency response by their physical properties and geometries. Piezoelectric materials used for these types of sensors include crystalline quartz and synthetic ceramics such as lead titanate and barium zirconate. Tourmaline crystal sensing elements were used for measurement of underwater blast pressure[143]. The design features for such transducers include tailoring of resonant frequency, which affects the mechanical design of the sensor housing, the geometry of the sensing element and the method by which the sensing element is interfaced to the sensor housing. Theoretically, small sized piezoelectric elements should be capable of extremely high resonant frequencies. However, as the size of piezoelectric element becomes smaller, it produces less charge at a given pressure. Since any charge produced must be shared between the capacitance of the element itself and the capacitance of attached cabling, output of the sensor at remote charge amplifiers falls as the sensor becomes smaller. This contradictory behaviour of response and accuracy has to be resolved.

---

[143] G. Bocksteiner, Evaluation of Underwater Explosive Performance PBXW-115 (AUST), DSTO Aeronautical and Maritime Research Laboratory, Australia, DSTO-TR-0297, Feb 1996.

For this, polymer film based piezoelectric materials are developed through polling. Piezoelectric polymer film gages produce electric charge under pressure. Several types of piezoelectric film material exists like nylon 3,5,7,11, polyvinylidene fluoride (PVF_2), and a co-polymer known as VF_2/VF_3. These materials are made piezoelectric by stretching the material and then polling it. Polling is exposing the material film to a slowly decaying intense electric field that realigns its molecules. Reproducible gage response and proper calibration make system ready for further use. There are several advantages of polymer film gages over piezoelectric crystals.

- They have low quality factor (the ratio of the resonant frequency to the bandwidth). Low quality factor corresponds to higher band-width. So, Piezoelectric film can be employed as a transducer at frequencies from DC up to GHz.
- They have flat frequency response over a wide range. This eliminates the self-ringing found in brittle crystalline materials. By varying the thickness of the film, the resonance can be brought down to low MHz for very thick films (1mm) and up to GHz for very thin films (1 micrometer).
- The capacitance is high enough for effective use at high frequencies when the signal from the piezoelectric film element is fed to a low impedance instrument.
- Piezoelectric polymer films such as PVF_2 also have pyroelectric properties (electric polarization induced by thermal absorption in crystals, with the polarization being proportional to the thermal change). The separation of responses of film due to shock heating of the sensor and the thermal pulse of the explosive is possible by such properties. This pyroelectric property of PVF_2 is important since adiabatic shock heating will occur in coincidence with the change in pressure it experiences. If the temperature changes in PVF_2 and its surroundings are identical, then there will be no heat transfer and the pyroelectric response will be due to adiabatic heating of the PVF_2 only.

Polyvinylfluoride based pressure gauge are also used for the measurement of detonation pressures. Its monomer units

($CH_2 - CF_2$) have 50% crystalline and 50% amorphous nature. This is a semi-crystalline piezoelectric material and under dynamic compression, electric charge is created at its poles. There is no need of any external power supply. A thin film of polyvinylfluoride is placed between two copper or gold platines. The gauge can be made thinner to enhance its sensitivity.

The other type of sensor element developed for blast pressure measurement is piezoresistive in nature, working on the wheatstone bridge set-up. It has a resistor mounted on a diaphragm. When pressure difference acts across diaphragm, diaphragm flexes and generates strain in the resistor. The resistance of resistor changes and some output voltage is generated. The support of diaphragm must be rigid enough for the pressure applied. Silicon is used as diaphragm as well as substrate material. Due to limited strength to weight ratio for silicon, the upper resonant frequency is limited to 1-6 MHz. Gage factor of such sensors is defined as ratio of fractional change in resistance to axial strain and is a measure of sensitivity of the sensor. Semiconductor piezoresistive gages have high sensitivity.

For the measurement of detonation pressure, Manganin gauge is generally employed. Manganin is an alloy of copper, manganese, nickel, iron and silica. The gages are composed of thin metal foil grid laminated to a plastic insulator similar to a strain gage. It has very low hysteresis and change in electric resistance due to temperature-change is also negligible. Resistance of this material changes under the action of dynamic pressure. It has low coefficient of sensitivity to pressure and is suitable for measurement of high pressure to the order obtained in detonation process. The change in electric resistance and applied pressure is found to be linear for the material. Typical manganin sensitivity is 0.003 ohms/ Kilobar. There are certain limitations of Manganin piezoresistive sesor.

- These gages are not made for resolution below pressure of 1 Kilobar.
- The gages are accurate only when forces act perpendicular to the gage face.

- To avoid gage heating manganin gages need to be activated using a pulsed power supply - applied just before pressure is experienced.

However, Manganin gages have measured pressures up to 400 Kilobars. These gages are typically used to measure detonation pressures internal to the explosive.

The sensors based on photoelastic effects can be useful for measurement of blast pressure of explosives. The photoelastic effect is due to the double refraction in transparent materials when a stress is applied. An explosive blast pressure wave applies uniaxial strain field on the photoelastic material in the direction perpendicular to the direction of light traveling through the material. When stress is applied, on a transparent material, in any direction, the index of refraction, increases along that direction, while the index of refraction, along the perpendicular direction, remains constant. This doubly refractive, or birefringent material, separates an incident beam of light into two diverging rays known as ordinary and extraordinary rays. The extraordinary ray propagates along the same direction at a slower speed than the ordinary ray traveling in perpendicular direction. Consequently, the higher the index of refraction of a material the slower the velocity of the extraordinary ray propagating through it becomes. Hence, one direction is known as the slow axis and the perpendicular direction is known as the fast axis. Experimental results on resonant frequency, pressure range, and response time show that the photoelastic based sensor is well suited to the measurements of blast waves.

For development of any new sensor, laboratory scale experiments are needed to understand the response time, pressure levels, resolution, sensitivity, linearity, repeatability and compatibility of sensors and gages. For this shock tubes are developed in which small scale experiments are conducted at lab level and the gages are qualified for field trials. In fact, attempts were made to measure explosive parameters based on simple and cost-effective principles. In one such attempt, dynamic impulse pressure measurement was attempted by vertical flexible cantilever type stem supporting a cylindrical mallet head[144]. The vertical rod was flexibly attached through an aluminium hinge, which deflects, when blast arrives at that point. The hinge undergoes a elastic, followed by plastic deformation and extent of deformation was treated as measure of blast pressure. Although, the gages measured pressure and impulse correctly, but they were damaged in the experiment, probably, due to transverse component of the impulse. In same experiment, blast pressure was also recorded by diaphragm type blast pressure gages. The principle behind such gages was location of initiation of rupture of diaphragm. Generally, these gages have a thin aluminium foil stretched in rigid support structure and governing parameters are diameter and thickness. When blast pressure is strong, the foil rupture at the periphery (edges), while for small pressures, ruptures starts from the center in radial outward direction.

Another simple method was based on the optical measurement of flight of a container lid under explosive action by a camera. The set up consists of a 30 mm thick rolled steel cylindrical vessel of diameter 1.5 m. The cylinder has a length-to-diameter ratio is kept 1. Hoops of 30 mm thick steel was welded onto the outside of the cylinder, with approximately 100 mm spacing between the hoops, to strengthen the container. The bottom of the cylinder was left open and the cylinder was suspended on four legs 300 mm above the ground to allow venting and access for hanging of the charge from the lid. A heavy circular plate (lid) was placed symmetrically on top of the cylinder with a calibrated mast fastened onto the plate in such a way that it would move with the plate when it is launched upward by the detonation. Measuring ports for the pressure sensors exist at 90° intervals in the circumference at three axial positions on the cylinder. The setup was named vertically launched impulse plate (VLIP)[145]. The charges were hung from

---

[144] H.S. Howe, The Measurement of Air Blast Parameters on the MINOR UNCLE Explosion Event Using Passive Instrumentation, DSTO Aeronautical and Maritime Research Laboratory, Australia, DSTO-TN-0035, March 1996.

[145] F.J. Mostert, CE du Toit, Measuring the Blast Output of Aluminized Explosive Charges in a Semi-Confined Environment, 2010 Insensitive Munitions & Energetic Materials Technology Symposium 11-14 October 2010, Munich, Germany.

the lid in a wire cradle in the centre of the apparatus in such a way that it was on the symmetry axis of the VLIP apparatus as well as in the middle of the cylinder. Reliable results were obtained during tests with aluminized charges containing varying percentages of aluminium. The method showed that, for the charges used, significant gain in impulse can be obtained if equal volume charges are considered. However, the mass weighted impulse of the aluminized charges is not significantly higher than that of the charges containing no aluminium.

Determination of detonation temperature is another decisive task. This helps in kinetics study of the explosive decomposition. It also helps in estimation of detonation products. Use of electro-optical methods, analysis of emission spectrum of detonation front, photoelectric scanning spectrometer etc has given some insight into the numerical values of detonation temperature, but accuracy and repeatability of results is still elusive. Since radiation nature of detonation or explosion is rather complex and it depends on temperature of surrounding atmosphere also, reliable determination of detonation temperature is not possible. 2-colour optical pyrometer is one such method, where radiation principles from a monochromatic light are exploited for determination of detonation temperature. In this case radiation energy of two near equal wavelengths of monochromatic radiation is measured. Using Wien's displacement law, the temperature is calculated.

Attempts are made to standardize various explosive test methods so that every explosive tests in different laboratories of the world result in same properties. International Study group fro the standardization of the methods of testing explosives was founded and several important recommendations are implemented[146]. The report advocates British system of measurement and converting numerical values to SI unit, fractional dimensions are indicated at certain locations. The report also gives details for detonation transmission, standardization of PETN

[146] H. Ahrens, International Study Group for the Standardization of the Methods of Testing Explosives – Recommendations Prepared up to Now by the Study Group", *Propellants Explosives Pyrotechnics*, Vol. 1, No ½, March 1977, p. 7–20.

detonator, imitation strength of electric and non-electric detonators etc.

Ballistic mortar test described in previous chapter has a steel mortar closed by a projectile and a long pendulum arm from a knife-edge suspends it. When detonation forces projectile out of the mortar, the mortar swings by an angle, which is proportional to explosive strength. The mortar weight is standardized to 311 kg and whole system weight is restricted to 470 kg. The projectile chamber in the projectile has a diameter of 123.8 mm and a depth of 127 mm. A coaxial chamber of 50.8 mm diameter and 139.7 mm length is used as firing or explosion chamber. The tolerance between projectile and its container is also indicated. The method of preparation of charge, assembly of components, fuze, initiation and measurement of end parameters are also specified. The swing angle is measured directly from the scale. The formula for weight strength relative to an standard explosive charge is calculated by formula given below. If TNT is sued as standard explosive, it gives TNT equivalent.

TNT Equivalent = $100 (1 - \cos\alpha)/(1 - \cos\alpha TNT)\%$.

Another method of measurement of explosive strength is Lead block test, where expansion due to explosion in cavity volume in a Lead block is ascertained to assess strength of the explosive. The 200 mm diameter × 200 mm long Lead block has a cavity of 25 mm diameter and 125 mm depth. The original volume of the cavity is taken as 63 ml. Expansion of cavity is measured by taking volume of water needed to fill the cavity, in a graduated measuring cylinder. It is not TNT, sometimes PETN/Potassium chloride charge is also used to give PETN equivalent. The charge density is 1.40 g/cc and with 10 ml explosive, the net expansion volume of cavity must be within 287 ml to 300-ml. This is used as calibration and after assuring correctness of lead block material and elastic properties for the standard explosive sample, test of new explosive is carried out. The correction for rise and fall in temperature during test is also made for the values.

Another important explosive property is Brisance, which is shattering power of the explosive and is measured by Modified

Kast method. This method is not described earlier in TNT equivalent or measurement of explosive sections. In this case deformation of copper crusher by detonation of explosive charge of fixed diameter and length is measured. TNT may be used as reference in this test. Zinc tube of 0.3 mm wall thickness, internal diameter 21 mm and height 80 mm is used for the preparation of the charges. The tube is filled with explosive at normal cartridge density to the height of 80 mm. Explosive is fired by initiator and booster of suitable strength. Booster is also a cylindrical charge of 21 mm diameter and lower height and is placed coaxially with the zinc tube. The initiator is also centrally placed over the booster. The apparatus placed over 20 mm thick steel plate, which in turn rests on a rigid base. The base is placed on a piston, which moves in a guide cylinder. The crusher is also cylindrical in shape and has 7 mm diameter and 10.5 mm height. The compression of crusher is measured by detonation of explosive. The crushing measured in length is converted into units of crushing by Haid and Selle Table. Minimum six reading are taken and averaged out to get value of crushing strength.

Measurement of detonation wave is another important area, which is investigated through many experimental means. In most of the cases, detonation wave is assumed stationary and unreacted explosives move through this and get converted into detonation products. Detonation wave has mainly three zones. The unreacted explosive faces shock front, which is also called chemical spike or von Neumann spike. Immediately behind this is steady chemical reaction zone. This is followed by Taylor expansion zone, which finally merges into detonation product zone. The separation of chemical reaction zone and unreacted explosive is called von Neumann spike (as told earlier) and the interface of chemical reaction zone with Taylor expansion zone is called Chapman-Jouguet plane. Out of various time and distance dimensions, pressure at C-J plane is one of the important parameters. This is determined empirically and theoretically in previous chapters, but experimental determination needs specialized instrumentations.

Flying plate test method is an important method to experimentally determine this parameter. In this test detonation wave (from an explosive charge) is applied on a metal plate, which is in contact with the explosive charge. Other side of plate is air. Detonation wave moves in explosive at certain velocity and at the interface, the wave is converted to shock wave in metal plate with different velocity. The free surface of the plate moves with another velocity. The shock pressure attenuates in the plate and as thickness increases, pressure continues to decrease. On a pressure-distance plot, zero thickness indicates pressure in the von Neumann spike. The kink in the pressure-distance plot corresponds to C-J pressure. The practical measurement is confined to measurement of velocity of free surface of the plate. It can be done by electro-contact type or velocity probe or oscilloscope technique or by optical means. For measurement using electro-contact probe, many probes with terminals located at stand-off for sequential closure. First pair of probes is placed at the explosive-metal interface to mark shock wave entry in the metal plate. The second pair of probes is placed at the free surface indicating time of exit of shock wave from the metal surface. Metal plate thickness (d) can be used in conjunction with measured time–difference (t_m) to give velocity of shock wave in the metal ($U_m = d/t_m$). A third pair of probe is placed at a known stand-off (h) to mark the time (t_h) for the movement of free end of the metal plate. Velocity of free surface of the metal plate can be obtained ($V_{fs} = h/t_h$). The experimental setting is made such that $t_h < t_m$. A series of experiments are carried out to get plate velocity.

Deflagration to detonation transition (DDT) is the behaviour of various explosive compositions, which is studied to understand safety parameters. It starts with accelerated burning leading to a series of compressive wave, which coalesce to form a shock of sufficient strength to initiate detonation. It has extremely complex mechanism and the dominant parameters controlling this are permeability and confinement. Permeability permits flow of hot gases through the explosive charge and is determined by pore size, shape and distribution. Particle size of explosive and compaction are controlled to arrive at positive features for permeability. Permeability decreases with decrease in particle size.

Special apparatus, which is a seamless tube with heavy end closures is used to measure it. It is filled with B-KNO_3 igniter to column length (~6.3 mm) and the explosive charge (column length ~ 295.4 mm). It has ionization probe (IP) and strain gages (SG) to monitor ionization fronts and internal pressure. Pressure is monitored at 4-5 locations. Relative time to detonation is monitored by IP, while interval between formation of first compressive front and the onset of detonation if measured by SG. A third parameter called pre-detonation column length of the explosive is also used for characterization of DDT. The predetonation length reaches a minimum with increase in particle size and then it increases again. In fact such 'U'-shaped curve is obtained for most of the explosives. Variation of pre-detonation length with %TMD also show similar trend[147].

Calculation of TNT equivalent on the basis of peak blast overpressure requires experiments with more safety precaution, more quantity of explosives and more hazards. Rather than using kilogram level of explosives, gram level explosive was proposed[148]. These experiments were safer and cheaper than the equivalent full-scale tests. The optical methods and high speed digital imaging used, allow more data to be recorded and more useful information to be derived than in typical full-scale outdoor explosive tests. These small-scale tests cannot fully replace fullscale tests, but they show useful potential. Inexpensive smallscale tests could be used to determine key experimental parameters and to bracket results to within an order of magnitude. Following gram-range testing, a limited set of full-scale tests should still be conducted to check the scaled results. It is also indicated that a single TNT equivalence value is insufficient to define explosive strength parameters, as observed previously with the full-scale tests. Shock wave Mach number and overpressure duration, as functions of radius from the explosion center, are required to properly define all parameters.

---

[147] D. Price, R.R. Bernecker, "Effect of Initial Particle Size on the DDT of Pressed Solid Explosives". *Propellants Explosives Pyrotechnics*, Vol. 6, No. 1, 1981, p. 5–10.

[148] Michael J. Hargather Gary S. Settles, Optical Measurement and Scaling of Blasts from Gram-Range Explosive Charges, *Shock Waves*, Vol. 17, 2007, pp. 215–223.

Advanced technique was employed to generate internal detonation velocity in the explosive by Chirped fiber-optic Bragg grating (CFBG) sensors[149]. The report explained progress on an embedded velocity diagnostic using a 125 micron diameter optical fiber containing a chirped fiber Bragg grating. As the chirped fiber Bragg grating was consumed by the moving detonation wave, the physical length of the unconsumed Bragg grating was monitored with a fast InGaAs photodiode. Experimental details of the associated equipment and data in the form of continuous detonation velocity records within PBX-9502 were also presented in the report. This small diameter fiber sensor had the potential to measure internal detonation velocities on the order of 10 mm/μsec along path lengths tens of millimeters long. Measurement of physical length and location of CFBG in the optical fiber became a critical requirement for the high explosives applications. For this purpose, optical backscatter reflectometer (OBR) was used. To get a higher resolution, a hot-tip micro-probe was used to locally heat the CFBG at selected locations along its length while observing the CFBG spectrum with a spectrometer. This method can be used to determine the length and location of the CFBG to about 250 μm resolution. The hot-tip probe method created local heating and caused a local red-shift of the reflected light at the exact position where the heated tip touched the grating. This local heating appeared as a dimple in the CFBG spectrum. Using horizontal translation stage, it was relatively easy to create an accurate map of wavelength vs. physical length. The set-up was further augmented with a femtosecond laser, which cuts very precisely with minimal heating damage adjacent to the ablation zone. The physical length and location of a "sacrificial" CFBG by laser-cutting the CFBG in 1 millimeter increments was mapped while observing the spectrum of the CFBG in real time. A proportional relationship between the bulk reflection of a CFBG and its length was observed and measured during the cutting operation. The test set-up was validated using PBX-9502. Raw pin data for standard explosive PBX-9502 gave average detonation velocity as 7.65 mm/μsec, which is very near to the reported and calculated values.

---

[149] J. Benterou, C. V. Bennett, G. Cole, D. E. Hare, C. May, E. Udd, Internal Detonation Velocity Measurements Inside High Explosives, Lawrence Livermore National Security Report No. LLNL-PROC-409969, 20 Jan 2009.

10

Geometrical Aspects in Explosives

All explosive charges are made in some geometrical shape and performance is also said to be dependent on geometry of the charges. Length to diameter ratio of cylindrical charges invariably affects blast pressures. Practical determination of density needs consideration of geometry and calculation is carried out by measurement of dimension and weight. Dimension of explosive charges are used for calculation of volume of the charge. Estimation of quantity of explosive for a given shape also desires some treatment of geometry of the charges. Although geometry plays a major role in all types of explosive charges, but they are significant from manufacture of pressed explosive formulation mainly. During pressing many different shapes are obtained and calculation of volume of the recess becomes an important task. In this direction, various geometrical shapes are analyzed and strategy for calculation of their volume is given in this part of the handbook. This type of calculations is helpful in calculation of density of the manufactured charges. In general, explosive charges are available in right circular cylinder or conical geometries. For standard geometries volume are given by wellknown formulae.

Cylinder (diameter 'D' and height 'h') = $\pi D^2h/4$

Cone (diameter 'D' and height 'h') = $\pi D^2h/12$

Cube (sides 'a', 'b' and 'c') = abc

Hemi-sphere (radius a) = $2 \pi a^3/3$

Cylinder

Solid cylinders are  characterized by two-dimensional parameters namely diameter (expressed as d) and height (h). Cross-sectional area of the charge is given by $\pi d^2/4$ and volume of the charge is given by $\pi d^2 h/4$.

For a cylindrical charge of diameter 20 mm and height 10 mm, area of cross-section is $\pi \times 20^2/4$ mm^2 = 314.16 mm^2 = 3.1416 cm^2. The volume of the charge is $\pi \times 20^2 \times 10/4$ mm^3 = 3141.6 mm^3 = 3.1416 cc. If weight of the charge is 5.35 g, then density of the charge is (5.35/3.1416 g/cc =) 1.70 g/cc.

Tubular

A variation in solid cylindrical charge comes, if it is tubular. In addition to external diameter, another internal diameter also becomes one of the significant parameter. If external and internal diameters are depicted by D and d respectively, then

Area of cross-section = $\pi(D^2 - d^2)/4$

Volume of the charge = $\pi(D^2 - d^2) \times h/4$

The above equation reduces to equations of cylindrical charges, if internal diameter is replaced with zero. For a tubular charge with external diameter, internal diameter and height as 30 mm, 15 mm and 20 mm,

Area of cross-section = $\pi(30^2 - 15^2)/4$ = 530.143 mm^2 = 5.30143 cm^2

Volume of the charge = $\pi\,(30^2 - 15^2) \times 20/4$ = 10602.87 mm^3 = 10.603 cc

If weight of the tubular charge is measured using weighing balance is 17.5 g, then density of the charge is (17.5/10.603 g/cc =) 1.65 g/cc.

Conical

Another type of variation in cylindrical charges is frustum of a cone. For a conical section characterized by base diameter 'D' and height 'h', volume is given by $\pi D^2 h/12$. Frustum of a cone has cone-tip chopped off by a plane parallel to the base. This is characterized by two diameters each valid for opposite ends and height. The diagram is shown in Figure 9 along with salient features. Bigger diameter is D, smaller end diameter is d and height of the frustum is h. Frustum of the cone is extended as shown by the dotted line to complete the cone. The extra length needed for the same is s.

For calculation of volume of the frustum of a cone, volume of full cone with diameter of base as D and height as (h + s) can be taken. From this volume, volume of another cone with base diameter d and height s can be subtracted to get volume of the frustum of the cone.

Volume of bigger cone = $1/12\ \pi D^2 \times (h + s)$

Volume of extended cone = $1/12\ \pi d^2 \times s$

In addition to this, from similarity of triangles, $h + s/s = D/d$. or, $h/s = (D - d)/d$. or $s\,(D - d) = h \times d$

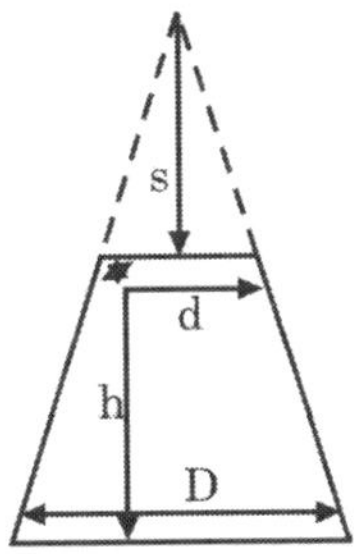

Fig. 9: Frustum of a Cone

Volume of frustum of the cone = $1/12\ \pi[D^2 \times (h + s) - d^2 \times s]$
$= 1/12\ \pi[D^2h + s(D^2 - d^2)] = 1/12\ \pi[D^2h + h\,d(D + d)]$
$= 1/12\ \pi\ h[D^2 + d \times D + d^2]$

It is worth noting here that calculation of volume based on average diameter may lead to errors. If instead of above mentioned equation, average diameter for the frustum of a cone is taken as (D+d)/2, and formula for cylindrical volume is used, then

Volume of frustum of a cone = $\pi [(D + d)/2]^2 \times h/4 = \pi h/16 \times (D^2 + 2dD + d^2)$. *This formulation is erroneous and it should not be used for calculation of volume of frustum of a cone.*

For illustration, a frustum of cone with diameters 30 mm and 15 mm along with height 20 mm is considered. From exact formulation,

Volume of frustum of the cone

$$= 1/12\ \pi\ 20\ [30^2 + 15 \times 30 + 15^2]\ \text{mm}^3$$

$$= 8246.68\ \text{mm}^3$$

Average diameter of the frustum of cone can be taken as 22.5 (30 + 15/2) mm and using formula for right circular cylinder,

Volume of frustum of a cone

$$= \pi\ 20/16 \times (30^2 + 2 \times 15 \times 30 + 15^2)\ \text{mm}^3$$

$$= 7952.15\ \text{mm}^3.$$

This value of volume by considering average diameter of both ends of frustum of a cone is although easy to calculate but incorrect. Averaging of diameter leads to lower estimate of volume of frustum of the cone. In this case the percentage error in calculation is (–) 3.57%. To correlate both the volumes, a universally true statement is considered as $x + 1/x \geq 2$. Equality sign is valid, if $x = 1$, else the term is always more than 2. Since both end diameters of frustum of a cone are different the ratio d/D is always not equal to 1.

$(d/D) + (D/d) > 2$ – On simplification

or, $d^2 + D^2 > 2\ dD$ – Add 2dD on either side

or, $(d + D)^2 > 4\ dD$ – Split left hand side

or, $4(d + D)^2 - 3(d + D)^2 > 4\ dD$

– Same coefficients on same side

or, $4(d + D)^2 - 4\ dD > 3(d + D)^2$

– Division by coefficient

or, $[(d + D)^2 - dD]/3 > (d + D)^2/4$

– Multiply both sides by $\pi h/4$

or, $\pi h\ [(d^2 + D^2 + dD]/12 > \pi h\ (d + D)^2/16$

– Word statement

or, Exact volume > volume by assumption of cylinder

In fact, estimation of volume by taking average diameter and formula for volume of cylinder always results in lower volume than actual. The percentage error in such calculations can be estimated also.

Percentage error = $25 \times (D - d)^2/\ [(d + D)^2 - dD]\%$

For the given data, percentage error = $25 \times 15^2/[45^2 - 15 \times 30]$ = 3.57%.

Sometimes for a conical charge, in place of height, angle (α) formed at the vertex of the cone is given. In such situations, height (h) and diameter (D) can be related using this angle.

Tan $(\alpha/2) = D/2h$.

Volume of cone = $\pi D^2h/12 = \pi D^3/[24 \text{ Tan } (\alpha/2)]$

Spherical

Spherical section is given to base of some container for specific purpose. Full spherical volume is rarely encountered in explosive geometry and most of the time hemispherical base is provided in recess or cavities. Volume of a hemi-sphere is given by $2/3\ \pi r^3$, where r = radius of the hemi-sphere. But most of the time full hemi-sphere is not available. Volume of a section of sphere, which is obtained by two parallel planes in same hemisphere can give a generalized treatment to all the cavities that are encountered. Let us assume that the section of a hemi-sphere of radius 'a' has angles α and β subtended at center as shown in Fig. 10. For the coordinate system shown, equation of the circular section is $x^2 + y^2 = a^2$. As shown in the figure, a small section of width dx at a distance x from vertical axis is considered. From the solid of revolution principle, volume of the disc created by rotation of this small circular-section about horizontal axis is given by $\pi y^2 dx$. This is to be integrated between limits to get volume of the spherical cup.

$$\text{Volume of spherical cup} = \int_{\alpha}^{\beta} \pi y^2 dx$$

$$= \int_{\alpha}^{\beta} \pi(a^2 - x^2)dx = \pi(a^2x - x^3/3)\Big|_{\alpha}^{\beta}$$

$$= \pi\,[(a^3 \cos\beta - a^3 \cos^3\beta/3) - (a^3 \cos\alpha - a^3 \cos^3\alpha/3)]$$

$$= \pi\, a^3\, (\cos\beta - \cos^3\beta/3 - \cos\alpha + \cos^3\alpha/3)$$

This is a general relation. For validation, this is checked for extreme conditions. If $\alpha = 90°$ and $\beta = 0°$, pure hemi-sphere will result and the value of volume for such section from the derived formula is $\pi\, a^3\, (\cos 0 - \cos^3 0/3 - \cos 90 + \cos^3 90/3) = \pi\, a^3\, (1 - 1/3 - 0 + 0) = 2/3\, \pi\, a^3$. This matches with volume of the hemi-sphere.

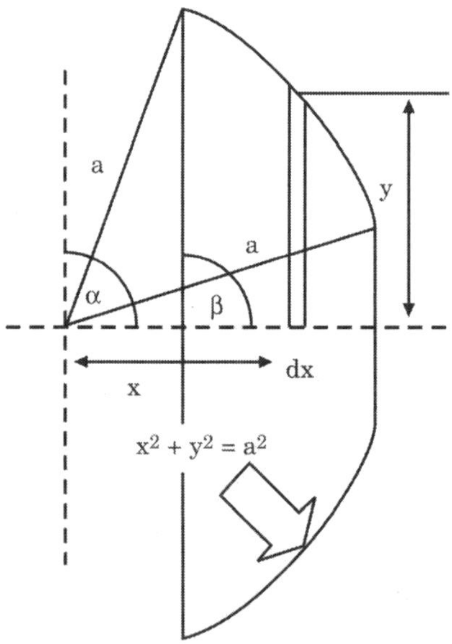

Fig. 10: Section of a Hemi-spherical Geometry

Another variation of such spherical curvature is observed in case of fillet radius provided at the corners of cylindrical containers. This is a different case as compared to the case depicted in Fig. 10. In the figure above, axis of rotation coincides with the center of sphere, but for filleted edges, axis of rotation is at an offset from the center of curved surface. In addition to this, the fillet is tangent to both the perpendicular surfaces and occluded angle for the corresponding circular section is 900 at the center. A typical situation is shown in Fig. 11. The corresponding circle has an equation $(x - b)^2 + y^2 = a^2$. The calculations steps are similar as explained above by taking a disc and obtain volume as a solid of revolution about the axis. In this case, fillet radius is taken as 'a' and straight part of the base has a radius of 'b'. A thin horizontal disc is considered, whose width is dy and radius is x. This can be integrated over vertical radius to get volume of the fillet-end.

$$\text{Volume of fillet end} = \int_0^a \pi x^2 dy = \int_0^a \pi[b - \sqrt{(a^2 - y^2)}]^2\, dy$$

$$= \int_0^a \pi[b^2 + a^2 - y^2 - 2b\sqrt{(a^2 - y^2)}]\, dy$$

$$= \pi\ [\{b^2 + a^2\}y - y^3/3 - a^2 b \operatorname{Sin}^{-1}(y/a) - by\sqrt{(a^2 - y^2)}]\ \Big|_0^a$$

$$= \pi\ [\{b^2 + a^2\}a - a^3/3 - a^2 b\pi/2]$$

$$= \pi a\ [b^2 + 2a^2/3 - \pi ab/2]$$

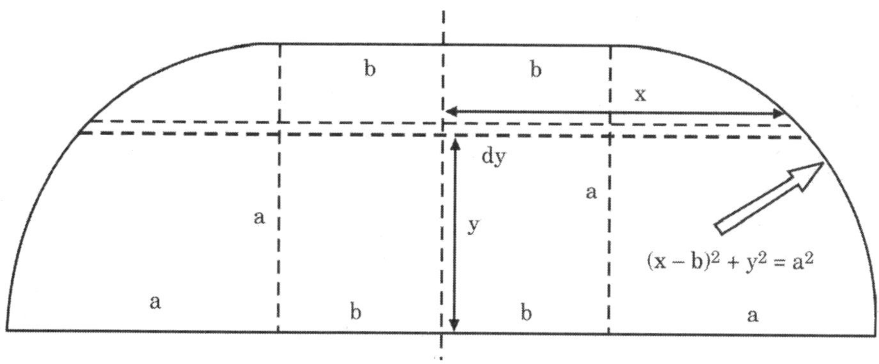

Fig. 11: Fillet End Geometry for volume calculation

For verification of the formula, geometry is made regular by changing its parameters. If b = 0, the geometry resembles a hemi-sphere with radius 'a'. From the derived formula, volume = $\pi a \times 2a^2/3 = 2\pi a^3/3$. This matches with the formula of volume for the sphere.

For a typical cylindrical geometry, cylinder diameter is 5 mm and total depth of cylinder is 4 mm. However, there is a fillet at one end with radius of 2 mm. If fillet radius is neglected, value of volume = $\pi 5^2 \times 4/4$ mm^3 = 25π mm^3 = 78.54 mm^3. However, exact calculation from the formula derived above gives, volume of the fillet-end cylinder to be made of two parts – (*i*) Volume for pure cylinder – height of cylindrical section will be only 2 mm as 2 mm portion is having curved edges. and (*ii*) volume of filleted section of cylinder.

Volume of parts 1 = $\pi 5^2 \times 2/4$ mm^3 = 12.5π mm^3 = 39.27 mm^3

For volume of part 2, a = 2 mm and b = 0.5.

Volume of part 2 = $\pi \times 2[0.5^2 + 2 \times 2^2/3 - \pi x^2 \times 0.5/2]$ mm^3 = 8.456 mm^3

Volume of fillet end cylinder = 39.27 + 8.456 mm^3 = 47.726 mm^3

Miscellaneous

In addition to the above-mentioned geometries, there are several other shapes, which were prevalent in explosives area. One such geometry is that of shaped charge, which can be treated as a combination of cylinder and cone. One end of the charge has a conical recess. By combination of formulation for the two, volume of the shaped charge explosive charges can be determined. If a shaped charge has diameter 50 mm, height 180 mm and conical recess with height 80 mm, then volume of calculated as difference of volume of cylinder and volume of recess.

Required volume

= $\pi \times 50^2 \times 180/4 - \pi \times 50^2 \times 80/12$ mm^3 = 301069.3 mm^3

= 301.0693 cc

If weight of such charge is 550 g, the density of the charge can be calculated as 1.82 g/cc.

Another type of geometry is found in rectangular explosive charge with conical recess spread along length. The geometry is shown in Fig. 12. In this case, volume is calculated as product of cross-sectional area with length of the charge. The crosssectional area can be thought of as made of two trapezoidal sections. Area of trapezoids is given by half of product of sum of parallel side and distance between parallel sides. This strategy is adopted and shown in Fig. 12.

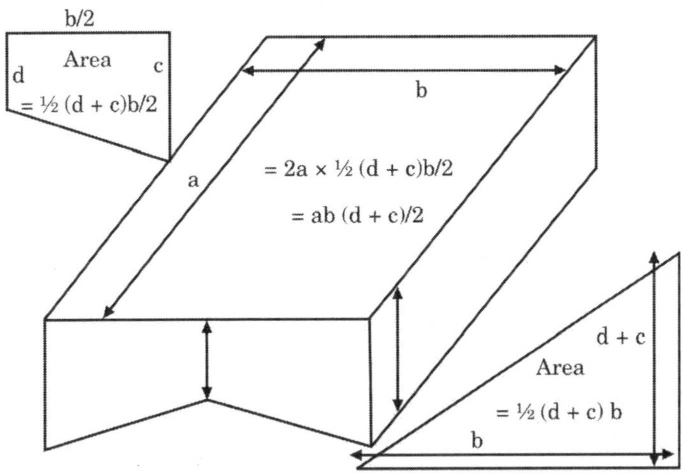

Fig. 12: Linear shaped charge configuration

Required volume = ½ ab (d+c)

This formula indicates that volume of such figure is equal to volume of a right-angled triangular cross-section of length 'b' and height (d + c) with length of charge equal to 'a'. The same is also depicted in Fig. 12.

11

Anomalies in Calculations

Although many calculations for explosives are illustrated in this book, but despite this, there are certain calculations strategies, which lead to certain anomalies in results. Such observations do not put any question mark on the developed mathematical formulations, but it only puts forward a caution in application of certain mathematical formulas. The anomalies in calculations may be in the form of variation in calculated and observed results or in the form of incorrect calculation results. Since experimental results depend on many parameters including method of measurement, measured quantity, measuring instruments, accuracy, repeatability, consistency and suitability of method etc, the first part of anomaly is generally offset in favor of either calculated or observed values. However, incorrect use of mathematical formulations may lead to many other types of erroneous calculations. This part of the book deals with such erroneous use of developed mathematical formulations.

Density is one of the simplest parameters relevant for the explosives. It is observed that density of a mixture is calculated from harmonic mean type of law and not as per arithmetic mean type of law. For example, let measured density of an explosive formulation is 1.70 g/cc. Aluminium (density = 2.70 g/cc) is added to this base composition such that ratio by mass of the explosive composition and aluminium is 60:40. As per theory developed earlier, theoretical maximum density (TMD) of the resulting composition is calculated below.

TMD of 60/40 (Explosive/Al) charge
= 100/((60/1.70) + (40/2.70)) g/cc = 1.99 g/cc

Howevcr, when charges are made, the density of charge by water displacement method can be found as 1.88 g/cc. This is much lower than expected density. This may seem like an anomaly in calculation and for resolving the matter quickly, sometimes arithmetic mean method is proposed. For same charge, calculation of density by arithmetic means (although not correct method) renders following results.

TMD of 60/40 (Explosive/Al) charge
= ((60 × 1.70) + (40 × 2.70))/100 g/cc = 2.10 g/cc

This result is further deviated from the experimental results and this approach for density calculation (by arithmetic means) is not correct. Calculation of density by taking harmonic means of explosive ingredients is to be used and it need not be doubted only because results are not matching to experimental observations. At the same time, experimental observations are also correct. The source of discrepancy is presence of unaccounted air-pocket or void volume fraction. *The harmonic means formulation for calculation of density is valid for calculation of theoretical maximum density (TMD) only and it need not simulate experimentally observed densities.*

If TMD of bare explosive is considered as 1.70 g/cc, then it can be concluded that addition of aluminium in explosive results in enhancement of porosity. Only (100 × 1.88/1.99 =) 94.47% of TMD (theoretical maximum density) is achieved in final composition practically. Every experimental density has certain porosity represented by percentage of TMD achieved. The lower density is due to presence of air pockets in the composition. On volume basis, the fraction of volume occupied by air is around 6.186 % for this composition. This value of volume fraction occupied by air cannot be calculated by any theoretical means and is purely a practical parameter. So, there is no discrepancy in experimental and mathematical determination of density. Both represent two different situations and percentage of mathematical density is always achieved as experimental density.

Now, if binder percentage in the composition is increased, the volume fraction of air reduces. In the aluminium-loaded composition, 5%, 10% and 15% wax (density = 0.9 g/cc) is added. Density of the charge is determined practically by water displacement method. Based on charge density of 1.88 g/cc for the aluminium loaded explosive composition, theoretical maximum density is also calculated.

For 5% wax charge, practical and theoretical densities are 1.83 g/cc and 1.78 g/cc (100/[(95/1.88) + (5/0.9)]) respectively. *Getting higher practical density than TMD is an anomalous result and is never possible.* Similar, calculation is done for 10% wax charge and practical and theoretical densities are 1.76 g/cc and 1.695 g/cc (100/[(90/1.88) + (10/0.9)]) respectively. For 15% wax charge, practical and theoretical densities are 1.67 g/cc and 1.62 g/cc (100/[(85/1.88) + (15/0.9)]) respectively. There is no problem with the calculation here, but the input data were not taken correctly. In fact, density of aluminium-loaded explosive composition as 1.88 g/cc is along with some volume of air also. However, after melting and reconsolidation with certain percentage of binder the volume fraction of air will reduce. So, in the calculation rather than observed density, TMD must be considered. In place of 1.88 g/cc as density of aluminium loaded explosive formulation, 1.99 g/cc must be considered. With 1.99 g/cc as density of base composition, loading 5%, 10% and 15% wax respectively results in densities of 1.88 g/cc, 1.78 g/cc and 1.69 g/cc. The practically observed densities are around 98.8% of TMD and volume fraction of air in the formulated composition reduces to 1.1% approximately. *The reduction in volume fraction of air from 6.186% to 1.1% results in anomalous results.* It must be kept in mind that while calculating density theoretically, all theoretical values must be considered, otherwise practical density more than TMD may be erroneously calculated. Since there is no method to predict volume fraction of air in the formulated composition, it is better not to take mixture of experimental and theoretical data in same calculation.

It is generally observed that at higher density a given explosive material exhibit higher detonation velocity. For example, RDX shows different detonation velocity at different densities.

Table 19 compiles some data for detonation velocity of RDX, presented in a paper[150].

Table 19: Variation of VOD of RDX with density

Density (g/cc)	***0.56***	***0.70***	***0.95***	***1.10***	***1.40***	***1.60***	***1.80***
Detonation velocity (m / s)	4050	4650	5800	6180	7460	8130	8750

Similar variations are observed for explosive compositions made of various molecules also. For example, Cyclotol 75/25 shows detonation velocities of 8300 m/s and 7950 m/s at densities of 1.76 g/cc and 1.62 g/cc respectively. However, if explosive ingredients are changed, higher densities need not necessarily exhibit higher detonation velocities. To show this anomaly, RDX/TNT 60/40 composition is considered. It is known that density of RDX and TNT are 1.806 g/cc and 1.654 g/cc respectively.

TMD of the formulation

= (60 + 40)/(60/1.806 + 40/1.654) g/cc = 1.742 g/cc

TATB/TNT 60/40 composition is considered for comparison. It is known that density of TATB and TNT are 1.938 g/cc and 1.654 g/cc respectively.

TMD of the formulation

= (60 + 40)/(60/1.938 + 40/1.654) g/cc = 1.813 g/cc

This clearly shows that TATB based formulation has higher density and obviously, it should give higher velocity of detonation. Calculation is carried out using Kamlet's method for both the formulations. At molecular weight using harmonic mean method is calculated. Equivalent molecular formula for molecular weight of 100 is calculated and number of moles of carbon, hydrogen, nitrogen and oxygen are raised so as to match the molecular weight. Oxygen balance is calculated. Heat release as discussed earlier is calculated and all the relevant parameters useful for calculation

150 M.H. Keshavartz, *Jrl of Hazardous Materials*, A121, 2005, p. 31–36.

as per Kamlet's algorithm, namely 'N', 'M' and 'ϕ' are calculated. From that detonation parameters are estimated and all the results are tabulated.

Parameters	*RDX/TNT 60/40*	*TATB/TNT 60/40*
Equivalent formula	$C_{4.58}H_{5.61}N_{4.82}O_{6.00}$	$C_{6.43}H_{5.57}N_{4.71}O_{6.00}$
Molecular weight	223.97	244.64
Density (g/cc)	1.742	1.813
Heat of formation (kcal/mol)	1.332	−11.213
Oxygen balance (%)	−42.54	−63.04
Kamlet parameters		
N	0.029499	0.026673
M	27.6405	27.6857
ϕ	5.800576	4.919396
Detonation velocity (m/s)	7951.1	7521.3
Gamma	3.005	3.070
Detonation pressure (kbar)	274.232	252.052

From the table given above, it is clear that despite higher density, TATB based formulation has lower detonation velocity. In addition to this, it is also observed that detonation pressure is higher for TATB based formulation. So, effect of density to some extent is visible in detonation pressure, but *higher density may not guarantee higher detonation velocity, if ingredients of the explosive formulations are changed. For same explosive ingredient, higher density gives higher detonation velocity.*

Detonation velocity and pressure are treated as governing detonation parameter, indicating performance of the explosive. The detonation performance of explosive can be measured by movement of metallic cylinder-walls. This sometimes contradicts the other detonation parameters like velocity of detonation or detonation pressure. Some similar examples are quoted below[151].

[151] Wu Xiong, "A Method for Evaluating the Detonation Performance and the Metal Accelerating Ability of Explosives", Proceedings of the International Symposium on Pyrotechnics and Explosives, Oct 12-15, 1987, Beijing, China, pp. 707–712.

Two HMX/TATB based explosive formulations are reported. First composition has density, velocity of detonation, detonation pressure and adiabatic exponent as 1.84 g/cc, 8300 m/s, 30.2 GPa and 3.2 respectively. For second composition, thc values are 1.838 g/cc, 7890 m/s, 29.5 GPa and 2.875 respectively. Although all the parameters are favourable for the first composition to be a better explosive, but metal accelerating ability in cylinder test is found better for the second composition. Another example is quoted for comparison of NQ and PETN. Both have comparable detonation velocity as 7980 m/s. However, NQ excels in all performance parameters. For NQ, density, heat of detonation, detonation pressure, temperature and adiabatic exponent are 1.626 g/cc, 3.673 kJ/g, 25.1 GPa, 1518 K and 3.15 respectively. For PETN, these values are 1.67 g/cc, 6.192 kJ/g, 30.0 GPa, 3018 K and 2.55 respectively. Rather than these detonation parameters, kinetic energy of cylinder wall at 6 mm and 19 mm displacements are used for measuring the energy or metal accelerating ability of explosives. The energy difference equation [$E - E0 = \frac{1}{2} \times (P + P0) \times (Vo - V)$] called Hugoniot energy equation derived in second chapter is widely used in the field of stationary detonation and this can be equated to specific kinetic energy ($1/2\ u^2$). The particle velocity is established to be more comprehensive detonation criteria for the explosive, which is correlated with other explosive properties.

$$\text{Gurney velocity, } \sqrt{2E}\ (\text{km/s}) = 1.736 + 0.4923 \times (u^2/2)$$

$$E^{6\ mm}\ (\text{kJ/g}) = 0.1766 + 0.249 \times (\rho \times u^2/2)$$

$$E^{19\ mm}\ (\text{kJ/g}) = 0.2847 + 0.30 \times (\rho \times u^2/2)$$

For RDX, specific kinetic energy is 2.41 kJ/g and experimental value of gurney velocity is 2.93 km/s. From the equation developed above, the gurney velocity is 2.92 km/s. For TNT, specific kinetic energy is 1.389 kJ/g and experimental gurney velocity is 2.44 km/s. From the equation reported above, calculated value is 2.42 km/s. Similar calculations are reported for HMX (2.615 kJ/g), PETN (2.31 kJ/g), Tetryl (1.636 kJ/g) for which experimental gurney energy are 2.97 km/s, 2.93 km/s, 2.50 km/s respectively. Using equation above and specific kinetic energy given in bracket, calculated gurney velocities are 3.02 km/s, 2.87 km/s and .54 km/s respectively. The calculations are repeated for cylinder test for 6 mm expansion. For HMX (1.894 g/cc), PETN (1.765 g/cc), TNT (1.63 g/cc), the experimental value of cylinder test for 6 mm expansion are 1.41 kJ/g, 1.255 kJ/g, 0.735 kJ/g respectively. The calculated values from the empirical relation are 1.41 kJ/g, 1.192 kJ/g and 0.74 kJ/g respectively. Similar values are reported fro 20 mm cylinder test for RDX (1.80 g/cc), HMX, PETN and TNT as 1.6 kJ/g, 1.745 kJ/g, 1.575 kJ/g and 0.975 kJ/g respectively. The calculated values from the above mentioned equations are 1.586 kJ/g, 1.77 kJ/g, 1.507 kJ/g and 0.964 kJ/g respectively. All the values are matching to experimental results. The specific kinetic energy can be considered as actual deliverable energy from an explosive rather than detonation velocity or detonation pressure, which are generated energy from the explosive.

One of reports discussing misuse of detonation velocity for the characterization of high explosive enumerates some anomalies through practical demonstration[152]. Three experiments are described to show anomalies and their explanations are also illustrated. In first experiment, a tube, thin enough to provide little confinement to the filled explosive, is filled with granular PETN loosely (density = 1.014 g/cc). VOD of the explosive is determined as 5.57 km/s. If the tube is now filled with similar form of explosive and water is filled in the tube (to the extent of 42.7% by volume) to fill the inter-granular spaces, another column of explosive is obtained. Now detonation velocity is determined again and it is found as 7.08 km/s. This gives anomalous conclusion that addition of water enhances energy of the explosives. This is contrary to the normal notion and explosive energy remains same in both the cases. To explain the observed effect, it is said that adding water decreases chemical energy release per unit weight of the explosive, but it has two positive effects. First it enhances density of composition because water replaces air. Second average mean molecular mass of detonation product reduces and number of moles of detonation products per unit weight of the explosive increases. Kamlet method can be used to justify the results. PETN

[152] M. Defourneaux, "About the Misuse of the Detonation Velocity for the Characterization of High Explosives", TR940060, NATO HQ, Brussels, Belgium, 18 March 1994.

($C_5H_8N_4O_{12}$) has density of 1.78 g/cc and molecular weight of 316. The heat of formation is –128.7 kcal/mol. The other relevant parameters are taken from Tables 2 and 4. Q = 1513.01 cal/g, N = 0.03163, M = 30.4, ϕ = 6.78306.

Velocity of detonation, D = 1.01 × √6.78306 × (1 + 1.3 × 1.014) km/s = 4.97 km/s. Detonation pressure, P = 15.58 × 1.014^2 × 6.78306 kbar = 108.659 kbar.

When water is added to the extent of 42.7% by volume, the arbitrary equation changes.

$C_5H_8N_4O_{12} + 0.427\ H_2O \rightarrow 2N_2 + 4.427\ H_2O + 4CO_2 + C$

Equivalent density, ρ = 1.427/[(1/1.78) + (0.427/1)] g/cc = 1.443 g/cc.

Chemical energy, Q = [–128.7 – 4x(–93.98 – 57.79)]/(316 + 0.427 × 18) kcal/g = 1477.9 cal/g

N = (2 + 4.427 + 4)/(316 + 0.427 × 18) = 0.0322133

M = 29.8922

ϕ = 0.0322133 × √(29.8922 × 1477.9) = 6.7707488

D = 1.01 × √6.7707488 × (1 + 1.3 × 1.443) km/s = 7.558 km/s.

P = 15.58 × 1.443^2 × 6.7707488 kbar = 219.653 kbar

Although chemical energy release (Q) and Kamlet parameter f both decreases by addition of water, increase in density causes increase in detonation velocity and detonation pressure of the experimental set-up by addition of water.

Second example given in the report is about stoichiometric hydrogen and oxygen mixture ($2H_2 + O_2$) at 20 K. experimentally it detonates with VOD of around 5 km/s. Now if concentration of hydrogen is increased, energy per unit mass reduces, because excess hydrogen behaves as diluents. However, increase in detonation velocity to the level of around 8 km/s is observed, if mixture is changed to ($9H_2 + O_2$). On similar lines as above, increasing hydrogen percentage in the composition, overall density of the mix reduces. Additionally the energy per unit mass of the explosive is also reduced due to dilution. Mean molecular weight of detonation products drastically comes down. However, number of moles of gaseous products of detonation increases and overall increase in detonation velocity may be justified on this ground.

Third example is about adding inert ingredient in explosive from 10% to as high as 75%. The additives are chemically, mechanically, and thermally inert. They may be asbestos, alumina, and other ceramics. Detonation velocity is monitored and it is observed that reduction is not very high. Detonation pressure is not found to be significantly changing by this process. It is stated that if density of additive and explosive are equal, then presence of additive increases detonation velocity. If additives have much higher density then velocity of detonation of resulting mixture reduces. For in between values, although chemical energy decreases, but detonation velocity is not affected significantly.

In fact compressibility of detonation products plays a major role in the detonation performance. If detonation products are incompressible (solid like in thermite), no energy is delivered by mechanical expansion of detonation products. So, requirement is to have highly compressible detonation products. In general, for CHNO explosives, reduction in carbon increases compressibility of detonation products. Compressibility is generally depicted by isentropic exponent (γ). If value of γ is large, products are less compressible. Increase in compressibility of detonation products reduces gamma. If incompressible additives are present, compressibility of detonation products reduces. This advocates research towards those molecules where detonation products have lowest proportion of solid phase like high-nitrogen-low carbon molecules. A probable explanation may be that diatomic nitrogen is less compressible at low pressures than tri-atomic water and carbon dioxide.

TNT equivalent is another figure of merit of explosives formulations, which is anomalous. The experimental determination of TNT equivalent is rather difficult and depending on theoretical calculations alone also leads to variations. Suppose peak blast pressure is considered as basis for calculation of TNT equivalent. The TNT equivalent is simply ratio of peak blast pressure at a given point. Practically it is observed that at different distances from the blast site, the ratio of peak blast pressures by trial explosive and TNT varies. The ratio is higher at higher pressures. In fact, thermo-chemical calculation also suffers from similar difficulties[153]. The calculations based on impulse, positive pressure phase, time duration etc also gives higher discrepancies.

TNT equivalent is through blast pressure measurement was deliberated in a paper highlighting the effects of charge geometry and point of initiation. The results were quoted for emulsion explosives and the governing parameters were enlisted as nature of explosive, effect, distance, shape and initiation. Although with explosive charges, scaling law was established and TNT-equivalent remains independent of the distance from explosion site, different TNT-equivalent were observed for blast overpressure (1.0) and impulse (0.7)[154]. The reason given for this anomalous behaviour was relatively less brisant emulsion explosive charge as compared to TNT, resulting in equivalent peak overpressure but reduction in performance over time for emulsion explosive. Brisance was not generally mentioned as a factor for TNT equivalent. Shape of explosive charge has a significant effect on TNT equivalent. As compared to spherical explosive charge, cylindrical explosive charge showed a higher blast overpressure in the direction perpendicular to the line of initiation. For large length to diameter (L/D) charge the rise of blast overpressure in perpendicular direction was more pronounced. If large and small L/D charges were fired, TNT equivalent was dependent on angular location with respect to initiation position. At 0° and 180°, TNT equivalent was high and at 45°/135°, the value of TNT equivalent was the least. Perpendicular to initiation point, the value of TNT equivalent was moderate. Large L/D (= 8.3) charge showed lower TNT equivalent at 0°, 45°, 135° and 180° than small L/D (= 1.0) charge. But at 90° orientation, reverse effect was recorded. Regarding point of initiation, pressure is more sensitive to this parameter than impulse.

---

[153] Zoran Bajiæ, Jovica Bogdanov, Radun Jeremiæ, Blast Effects Evaluation Using TNT Equivalent. *Scientific Technical Review*,Vol. LIX, No.3-4, 2009, pp. 50–53.

[154] Bart SIMOENS, Michel H. LEFEBVRE, Fumiyoshi MINAMI, Influence of Different Parameters on the TNT-Equivalent of an Explosion. *Central European Journal of Energetic Materials*, Vol. 8, No. 1, 2011, pp. 53–67

12

Future Directions

Despite underlying computational approaches adopted for performance prediction of explosives, the exact nature of effects of detonation is difficult to judge correctly. It is observed that each area of computation must be strengthened to get realistic estimates. The calculation strategy for physics and chemistry behind detonation is at crossroads and dilemma is about length and time scales to be adopted for calculations in explosive performance. In general average temperature of a city can be said to be 29°C, but there will be spatial variations in temperatures. On similar lines for performance of explosives, small porosities, voids and discontinuities may not affect the performance significantly, but for initiation, presence of such bulk irregularity may be important. The dilemma is about use of continuum mechanics codes or molecular dynamics codes for such calculations. Classical equations of motion may be useful and in quantum mechanics, empirical potentials become a vital tool for calculations.

Similarly initiation of explosive through hot spot mechanism has become an old system. There are manufacturing dependent performance variations, which must be considered. In addition to this, these microscopic modes of initiation may also be supplemented with macroscopic mode of initiation like friction and impact. The most important part of such calculations is the mesascale calculations, on which much attention is not paid. Initiation by shock, impact, flyer plates, shaped charge jet etc is always explained using hot-spot creation. However, further initiation schemes like visco-plastic flow and shear heating at meso-scale level may be investigated and mathematical formulation is still not available through these modes of explosive initiation.

The equation of state for reactants as well as products is also another area of concern and invariably, more accurate results are possible with more number of empirical constants. The method of calculation of these constants is also not standard and it varies from location to location. Without proper equation of state, even if product of explosion is arrived at by alternate schemes, the performance parameters remain illusive.

The unification of various theories is attempted in all areas of science. However, a single kinetic model for explosive initiation is not available at present. The low temperature initiation by cookoff, slow heating or certain intermediate endothermic reaction etc is presently handled by different approach, while high temperature initiation mechanisms like shock or single exothermic step etc has different coefficients in equations. Unified kinetic model for initiation of explosives is not available. In fact growth of reaction from thermal stimuli to the level of shock initiation has different time and length scales and assemblage of both together is still not possible. Direct initiation of a self-sustained detonation is studied numerically in a paper[155]. The entire chemical heat release process during detonation is simulated as three steps of chain reaction – initiation, branching and termination. The numerical simulation describes characteristic time for each of the three steps. Minimum allowable shock pressure for initiation is a function of characteristic time for the initiation step. When time increases minimum shock pressure I the decaying period becomes larger. For small values of characteristic time of initiation, new detonation wave is produced behind the leading shock in the decay period and amplifies the leading shock abruptly.

Like burning rate models in the area of propellants, where deflagration plays a major role and particle burning is controlling

---

[155] S.A. Hashmi, A. Fattahi, "Effect of the Chain Initiation Reaction Rate on Direct Initiation of Detonation". *Combustion Explosion and Shock Wave*, Vol. 46, No. 3, May-Jun 2010, pp. 325–331.

the combustion process, the burn models of various hydrocarbons must be included in explosion calculations also. Such particle burn approach is seldom used in explosives. In fact role of aluminium in explosive formulation is still not understood due to this very reason. Partial combustion of aluminium and partial heat release along with location of heat release becomes major concern for explosive calculations. The treatment of aluminium oxide as gaseous molecules or heat sink condensed molecule is also another concern to be addressed by modern calculation schemes. Overall, particle burn approach for detonation prediction needs to be developed.

The physics of shock waves is said to be developed long back but shock wave augmented by chemical heat release during detonation of high explosives is not fully developed. Chemical kinetics of heat release, heat feedback and molecular motion behind shock wave are some of the areas which need proper investigation and study. Based on theoretical and experimental results, motion of particles behind the shock wave is described by a physico-mathematical model. However drag coefficient of particles is proposed to be a complex function to match calculated results with experimental observations of particle trajectory and velocity. The unsteady motion of particles is also described[156]. However a fully theoretical model or calculation approach is still unavailable.

In fact simple detonation waves are not understood clearly, but human thought has devised methods to monitor and control the process of detonation by adding inert particles of tungsten, alumina and silica. The mechanism becomes more complex and using steady one-dimensional detonation wave, attempts are made to get numerical insight into the detonation suppression mechanism. The control parameters are channel diameter and detonation cell size and minimum mass of the particles and the characteristic cloud size is assumed necessary for such predictions. Particles with high specific heat and high phase-transition heat

---

[156] A.V. Fedorov, A.V. Shul'gin, S.V. Poplavski, "Motion of a Particle Behind the Shock Wave Front". *Combustion Explosion and Shock Wave*, Vol. 46, No. 2, Mar-Apr 2010, pp. 207–215.

(for melting) influence these phenomena significantly[157]. However this phenomenon is not completely understood. On one side, inert material like aluminium is added to enhance the blast performance of explosives formulations but being an inert material, this has detrimental effect on detonation phenomena itself.

Modelling at micro-scales as well as macro-scales is attempted, but direct calculation at meso-scale is not possible at present by the available approaches. Various hydrocodes are also developed for assessment of explosive devices and attempts are made to simulate the actual cycle of operations and effects.

Computational approaches require meshing of a given domain and it converts continuum into discrete domains. Micron sized meshes in the continuum hydrocodes can be used to simulate behaviour of energetic materials under shocked conditions. The interaction of crystals of explosive as and binder matrix is possible to simulate only when such approaches along with particle burn models are developed for explosive predictions[158].

Detonation does mechanical work on the environment by rapid release of chemical energy. Generally mass ratio of explosive charge to metal gives indication about acceleration of a metal plate by explosion products. If non-porous plates of aluminium, steel, copper and tantalum are driven by explosive of same mass ration, the acceleration may have little variation, but building same mass by layers of plates or using non-porous plates make the similarity vanish and popular understanding fails to explain the observation[159]. The observations cited in the paper may be trivial observations but explanation is unavailable.

---

[157] P.A. Fomin, J.R. Chen, "Effect of Chemically Inert Particles on Parameters and Suppression of Detonation in Gases". *Combustion Explosion and Shock Wave*, Vol. 45, No. 3, May-Jun 2009, pp. 303–313.

[158] P.J. Haskins, M.D. Cook, "Modelling of Initiation and Detonation", 24th Pyrotechnic Seminar, California, USA, 27-31 July 1998, pp. 749–756.

[159] I.M. Voskoboinikov, "Velocities of Acceleration of Plates Made of Different Materials by Detonation Products". *Combustion Explosion and Shock Waves*, Vol. 45, No. 3, May-Jun, 2009, pp. 342–347.

Development of micro-explosive systems is considered one of the enabling technologies with widespread applications. However, such applications need proper scaling of explosive effects. Scaled reduction by a factor of 100 to 1000 is possible to miniaturize explosive devices. The physical scaling is not sufficient to address this issue. A few meter dimension of charge can be physically reduced to some mm, but the reaction zone length and time should also be scaled. This means that main charge explosives, which are secondary explosives for large-scale systems should be replaced with primary explosives for small-scale system to simulate scaling of reaction zone. Replacement by such new explosive materials needs development of linear and nonlinear stability theory for non-ideal detonation that can incorporate non-ideal equation of state and realistic reaction rate laws for condensed explosives. For miniaturized secondary explosive charge, development of nano-engineered energetic materials and explosives is underway.

Another way to tackle such scaling problems is to understand transient detonation propagation mechanism, where miniaturization shifts a quasi-steady design paradigm. Microdetonics is an unexplored area, which has many potential applications but method for scaling has to be finalized. Microdetonics is dominated by transient effects, which are associated with initiation of explosives by small sources. Development of theory for fast and sensitive chemical kinetics, detonation and shock diffraction and their effects on detonation propagation are to be established. Study of corning turning (transient effect of the change in confinement of detonation), dead zone (shock uninitiated explosive), coupling of mechanics of phase transformation and magnetohydrodynamics are areas where microdetonics has to develop[160].

The effect of process parameters on sensitivity characteristics of explosive molecules is not established yet. In fact crystal morphology, particle size and particle shape parameters affect sensitivity but the reason and predictive approach is still missing.

---

[160] D. Scott Stewart, "Miniaturization of Explosive Technology and Microdetonics", Mechanics of 21st Century, ICTAM04.

However experimental evidences are produced. While studying friction sensitivity of ε-CL-20, obtained by recrytallization with different anti-solvents, it was observed that friction sensitivity of final products were different. With cyclohexane as anti-solvent, the friction sensitivity was 78 N, while with toluene or chloroform, the friction sensitivity was 110 N. During process of crystallization, if stirring speed is increased, particle size comes down and reduced sensitivity to friction (in N) is seen. These effects are difficult to model for given explosive molecules and prediction for new molecules becomes equally challenging.

One of the major challenges in the area of explosives is development of suitable diagnostic technique to capture detonation. The transient and extreme values of various parameters in detonation zone make all sensors, non-operational, posing a major handicap to the investigative research on explosives. Several attempts are made by researchers to capture some of the events during explosion and predict behaviour of explosion. However, these are at experimental level only without any consensus among the research on the method as well as measured parameters. Most of the efforts are in the area of using radiation for detonation zone studies and are based on typical instrument or set-up developed in a laboratory. Universal acceptance of such methods and standardization of the schemes is still a distant reality. In one such effort, synchrotron radiation is utilized for study of detonation in condensed high explosives[161]. It is claimed in the paper that values of parameters in the Neumann spike and at the Jouguet point are determined. This opens an entirely new domain in studying detonation processes. Density distribution on the detonation wave front and in the expanding products is determined. The unloading adiabat of detonation products of explosives can be reconstructed by the experimental results. In another effort multi-frame proton radiography was used to study fast hydrodynamic processes in static as well as dynamic

161 V.M. Titov, E.R. Pruuel, K.A. Ten, L.A. Lukyanchikov, L.A. Merzhievskii, B.P. Tolochko, V.V. Zhulanov, L.I. Shekhtman, "Experience of Using Synchrotron Radiation for Studying Detonation Processes". *Combustion Explosion and Shock Wave*, Vol. 47, No. 6, Nov–Dec 2011, pp. 615–626.

experiments. Initiation and propagation of detonation wave in condensed phases, shock wave propagation in inert materials, hydrodynamic instability in metals, dynamic deformation and fracture of liners and plates, pall fracture, formation of shaped charge jet and its interaction with obstacles etc are studied. The developed method is claimed to give high spatial resolution, multiframe recording, unlimited thickness of test specimen and dynamic range of recording are coted as major advantages of this method. The method is under further development at Russian Federal Nuclear Center – Institute of Experimental Physics (RFNCVNIEF), Sarov, Russia[162]. In another effort, microwave interferometers are used for diagnostics of shock waves and detonation processes. Many advantages of using microwaves are also enlisted. Since most of the explosives are radio transparent, direct diagnostics of initiation and propagation of detonation is possible using these waves. It is also highlighted that presence of conducting materials interferes with radio transparency of these explosives. Another advantage is derived from the longer wave length of microwaves. The surface roughness of the order of 1-100 microns (covering all gas-dynamic discontinuities and surface roughness) act as specular reflectors, while optical radiations are diffused from such surfaces. Microwave displays another distinguishing characteristic. Reflection of such waves occurs at a jump of permittivity (conductivity) of a medium. Thus, making it useful for diagnosis of density jump, as well as structural transformations like ionization, dissociation, condensation in shock waves and compression/rarefaction waves. The wide range of wavelengths available for microwaves put forth another major point for versatility of the method[163].

---

[162] V.V. Burtsev, A.I. Lebedev, A.L. Mikhailov, V.A. Ogorodnikov, O.V. Oreshkov, K.N. Panov, A.V. Rudnev, O.V. Svirskii, M.A. Syrunin, Yu.A. Trutnev, I.V. Khramov, "Use of Multiframe Proton Radiography to Investigate Fast Hydrodynamic Processes". *Combustion Explosion and Shock Wave*, Vol. 47, No. 6, Nov–Dec 2011, pp. 627–638.

[163] V.M. Bel'skii, A.L. Mikhailov, A.V. Rodionov, A.A. Sedov, "Microwave Diagnostics of Shock-Wave and Detonation Processes". *Combustion Explosion and Shock Wave*, Vol. 47, No. 6, Nov–Dec 2011, pp. 639–650.

One of the areas, which got considerable amount of renewed interest over first 10 years of 21st century is pulse detonation engine (PDE). This has potential for increased performance values on relatively low-cost engine architecture. In this case detonation waves are created repeatedly in a chamber, propagating through the fuel-air mixture. The discrete impulses give rise to quasi-steady thrust. The intermittent high pressure in the chamber produces thrust by the exiting momentum flux. The sequence of pulsing should be executed quickly to produce nearly constant thrust. Such engines offer many advantages over conventional rocket engines. Pulse detonation engine has high thermodynamic efficiency than any other known cycle. It can be operated by conventional as well as specific fuels used in aerospace propulsion like ethylene, Kerocene, JP10 etc. It does not have complicated parts like turbo-pump or compressor and construction is also simpler. Absence of moving parts makes it more robust. It has the advantages of ramjet engine but disadvantages like boosting speed to certain high cruise values (as needed for ramjets) are not present. Thrust amplification and thrust vector control is possible by providing several identical units in assembly.

At present very little experimental data exists for such systems, but theoretical and numerical modeling of the phenomena is a challenge to be addressed by further research. Many initial trials, experiments, theoretical analysis concentrated on a single aspect of such engine and is a major step towards realization of a workable engine. Brophy *et al.*[164] studied effect of fuel mass distribution on performance of pulse detonation engine. Both initiation characteristics and fuel based specific performance values were monitored. Frolov[165] designed and experimentally demonstrated liquid fueled air breathing pulse detonation engine. The operation of engine was stable and many interesting finding were reported.

---

164 C.M. Brophy, R.K. Hanson, "Fuel Distribution Effects on Pulse Detonation Engine Operation and Performance". *Journal of Propulsion and Power*, Vol. 22, No. 6, Nov–Dec 2006, pp. 1155–1161.

165 S.M. Frolov, "Liquid-Fueled, Air Breathing Pulse Detonation Engine Demonstrator: Operation Principles and Performance". *Journal of Propulsion and Power*, Vol. 22, No. 6, Nov-Dec 2006, pp. 1162–1169.

The initiation energy for detonation can be reduced by half for smooth walled straight tube without using any turbulent created by obstacles. The existence of detonation peninsular in pre-detonator is another important finding. Gas dynamic focusing of compression waves generated by accelerating flame is observed probably due to curvilinear reflecting surfaces. Initiation of detonation in spray of liquid fuel in air was demonstrated by successive triggering of two igniters. The demonstrator gave specific impulse of the order of 700-800 seconds. However mathematical formulations of various events were not understood. In another paper[166], continuous spin detonation in liquid-propellant rocket motors and ramjet combustors was investigated. The presence of a critical minimum size of combustor needed for sustaining detonation was experimentally shown. Simple geometrical shape like annular cylinders without duct constriction was suitable. Experimentally, theoretically and numerically, it was shown that transonic transition occurs in the flow during continuous spin detonation. An ideal continuous detonation process ensures a higher specific thrust of the chamber. The efficiency of air-breathing pulse detonation engine is evaluated by combined analytical and numerical method[167]. System configuration, operating parameters and flight conditions were varied. Unsteady gas-dynamics, detailed detonation propagation, flow expansion through nozzles was used in the analysis. The proposed engine outperforms its ramjet counterpart for all the flight conditions but net benefit decreases with increasing flight Mach number.

Mathematical treatment can help in effective discovery of energetic materials, by resorting to quantum chemical calculations. In one paper[168], heat of formation of known explosive molecules

---

[166] Fedor A. Bykovskii, Sergey A. Zhdan, Evgenii F. Vedernikov, "Continuous Spin Detonations". *Journal of Propulsion and Power*, Vol. 22, No. 6, Nov-Dec 2006, pp. 1204–1216.

[167] Fuhua Ma, Jeong-Yoel Choi, Vigor Yang, "Propulsive Performance of Airbreathing Pulse Detonation Engines". *Journal of Propulsion and Power*, Vol. 22, No. 6, Nov-Dec 2006, pp. 1188–1203.

[168] Lulu Huang, Lou Massa, Jermone Karle, "Discovery of Energetic Materials by a Theoretical Method (DEDTM). *International Journal of Energetic Materials and Chemical Propulsion*, Vol. 10, No. 1, 2011, pp. 33–44.

like CL-20, HMX, RDX, TNT and PETN are calculated by *ab initio* quantum chemical B3LYP/6-31G(d, p) calculations and heat of sublimation is calculated by an empirical packing energy formula. Both these quantities are used to calculate heat of formation of condensed explosives, which gave detonation performance by CHEETAH computer program, matching to their reported performance. The approach is stated to be useful for prediction of new explosive molecules. Such techniques are mandatory for future research and identification of new explosive molecules.

One of the major areas in the domain of explosive performance is underwater explosion. The area is unique due to incompressibility of water as compared to air. Investigations are possible in this domain by experimental means only, as theoretical predictions are difficult and virtually nonexistent[169]. For a given explosive charge, the impact on surrounding objects depends on the properties of the energy transfer medium. Water has more incompressibility than air, so it has much higher energy transfer efficiency. In the underwater explosion, explosion of explosives, results in formation of high pressure gas bubble, from which a primary shock wave travels in outward direction. This expands gas bubble till pressure drops below that of the surrounding. This is followed by compression of gas bubbles and so on gas bubble size expands and compresses alternately. At every expansion process, shock wave emanates. The complex phenomenon of energy transfer from an underwater explosion has not yet yielded a complete mathematical description. The details of blast shock waves cannot be estimated by hydro-codes, because of complexity of the explosion phenomenon and the purely numerical or analytical solutions do not exactly conform to experiments. Underwater explosion remains an area of future research, modeling and experimental validation.

The current level of mathematical simulation is also not adequate and it is desired to develop modern high resolution, multi-dimensional and multi-material, time-dependent simulation.

---

[169] E. Kowsarinia, Y Alizadeh, H.S. Salavati Pour, Experimental Evaluation of Blast Wave Parameters in Underwater Explosion of Hexogen Charges. *IJE Transactions B: Applications*, Vol. 25, No. 1, February 2012, pp. 65–72.

Suggested Books for Further Reading

1. Detonation of Condensed Explosives, J. Taylor, University Press, Oxford, UK 1952.
2. The Science of High Explosives, by Cook, M.A., New York/ Reinhold Pub. Corp., 1958.
3. Chemistry and Technology of Explosives, T. Urbanski, Vol. 1 to Vol. 4, Pergamon Press, UK, First Edition 1964, Current Reprint 1990.
4. Gasdynamic Theory of Detonation, Heinz D. Gruschka, Franz Wecken, Gordon and Breach Science Publishers, New York, 1971, ISBN 0-677-033702.
5. Introduction to Gasdynamics of Explosions, Antoni K. Oppenheim, Springer Verlag, New York, 1972, ISBN 0-387-81083-8.
6. The Science of Industrial Explosives, by Cook, M A, IRECO Chemicals/USA, 1974.
7. A Manual for the Prediction of Blast and Fragmentation Loading on Structure, W.E. Baker, J.J. Kulesz, P.S. Westine, P.A. Cox, J.S. Wilbeck, Report no DOE/TIC-11268, Nov 1980, US Department of Energy, Texas, National Technical Information Service (NTIS).
8. Explosive Shocks in Air, 2nd edition, Gilbert F. Kenney and Kenneth J. Graham, Springer-Verlag New York, 1985, ISBN 0-387-15147-8.
9. Explosives, Propellants and Pyrotechnics: Land warfare, Vol. 2, by Lee RG Ed, UK/Brassey/1989.

10. Modern Methods and Applications in Analysis of Explosives, by Yinon, Jehuda/Zitrin, Samuel, Chichester/John Wiley & Sons, 1993.

11. Detonation of Condensed Explosives, by Cheret, Roger,New York/Springer – Verlag, 1993.

12. Test Methods for Explosives, Muhamed Suceska, Springer-Verlag, New York, 1995, ISBN 0-387-94555-5.

13. Introduction to the Technology of Explosives, by Cooper, P.W./Kurowaki, S.R., New York/Wiley-Vch, 1996.

14. Explosives Engineering, By Cooper, Paul W., New York, US/VCH Publishers Inc, 1996, ISBN 1-56081-927-8.

15. Explosive Effects and Applications, by Zukas, J.A. Ed./ Walters, A.P. Ed., New York/Springer, 1998.

16. Adhunik Uchcha Visphotak (Modern High Explosives), by Singh, H./Shekhar, H., Pune, India, 2006.

17. The Chemistry of Explosives, by Akhavan, J., Cambridge/ Royal Society of Chemistry, 2006, ISBN 0-85404-640-2.

18. Explosives 6th, by Mayer, R./Kohler, J./Homburg, A., Weinheim/Wiley-Vch, 2007.

19. Numerical Modeling of Explosives and Propellants, by Mader, C.L., London/CRC Press, Taylor & Francis, London, UK, 2008.

Subject Index

S

T

V

W